S0-CBP-691

Blood Ties

Blood Ties

A story of falconry
and fatherhood

Ben Crane

The University of Chicago Press

The University of Chicago Press, Chicago 60637
© 2018 by Ben Crane
All rights reserved. No part of this book may be used or reproduced
in any manner whatsoever without written permission, except in
the case of brief quotations in critical articles and reviews. For more
information, contact the University of Chicago Press, 1427 E. 60th
St., Chicago, IL 60637.
Published 2020
Printed in the United States of America

29 28 27 26 25 24 23 22 21 20   1 2 3 4 5

ISBN-13: 978-0-226-71473-8 (paper)
ISBN-13: 978-0-226-71487-5 (e-book)
DOI: https://doi.org/10.7208/chicago/9780226714875.001.0001

This is an Apollo book, first published in the UK in 2018 by Head of
Zeus Ltd.

Library of Congress Cataloging-in-Publication Data

Names: Crane, Ben (Photographer), author.
Title: Blood ties : a story of falconry and fatherhood / Ben Crane.
Description: Chicago : University of Chicago Press, 2020.
Identifiers: LCCN 2019037140 | ISBN 9780226714738 (paperback) |
    ISBN 9780226714875 (ebook)
Subjects: LCSH: Crane, Ben (Photographer) | Falcons. | Human-
    animal relationships. | Falconry.
Classification: LCC QL696.F3 C735 2020 | DDC 598.9/6—dc23
LC record available at https://lccn.loc.gov/2019037140

♾ This paper meets the requirements of ANSI/NISO Z39.48-1992
(Permanence of Paper).

To Pete for the space,
to Steve and Hollie for the tools,
and E and J for the pant-stretching mirth

INTRODUCTION

The Falconer and the Hawks

I n the warm, womb-like space of the cottage, the light from the open fire flickers and casts dull shadows of birds across the wall. On my gloved hand is a slender, lightweight and beautifully patterned female sparrowhawk; to my left, a smaller but no less impressive male. Both hawks emanate a quiet, self-contained calm. A fine balance of delicate precision, coiled, unsparing instinct, contained within a gossamer skein of feather, skin, muscle and bone. They remind me of that thin slither moment just before a jack-in-the-box pops. Both hawks arrived from the wild, injured; to have them legally in my possession is a rare pleasure.

It is not commonly known, but hawks smell.

After her accident, the female sparrowhawk's breath became a mixture of metallic sour fish and ammonia; a rancid, tacky odour that remained on my skin and in my nose. The male, cooped up and cloistered from the world, had lost his lustre, had the stale rot of forced captivity.

All hawks have a protective sheen, a bloom, weather-proofing against the rain. A fresh bloom on a hawk with per-fect feathers is divine. The detailed rehabilitation of each hawk complete, they now emit a low, musty tang, the smell of soft earth, rotting peaches, the marmalade mossiness of dried twigs. Anything that smells this good is undeniably fit and ready to go free.

The rehabilitation and release of these sparrowhawks is the penultimate stage and culmination of an obsessive journey – a journey that sent me in search of a visceral and unmediated relationship with the natural world. One that exposed me to many profoundly moving moments. I witnessed golden eagles soaring from monolithic granite mountainsides, over Austrian castles and across the blizzard-streaked landscapes of Germany and northern Europe. In temperatures of minus 32°C, near the Sioux Indian reservations of South Dakota, a tiny speck, a trained falcon, looped over, stooping hundreds of feet. At extreme velocity she sliced across a pheasant, sending it wheeling and tumbling to the ground, as descending blood froze in beads and pomegranate-red flesh was cast across snow. In the hazy heat of a Croatian summer dawn, I saw the speeding smear of a wild sparrowhawk chasing coturnix quail into an azure sky, scattering them in a mottled brown explosion of little clockwork fireworks. In ripping cross winds, and the harsh winter of South Dakota, I witnessed the most powerful falcon of all. A rare, wild, almost black, melanistic gyr falcon, a silhouette of wraith-like proportions, set against the reflective silver ripples of a vast lake. In Texas I followed a family unit of wild Harris hawks. Shrewd and cunning pack hunters, they scudded after rabbits through the desert scrub where pale sand meets turbulent sea on the Gulf of Mexico. It was also in Texas that I trapped a diminutive American kestrel. Shoulders cobalt, tail feathers metallic burnished bronze, he was brighter than a hummingbird and bit me hard before being released.

Of all the journeys I have undertaken it was time spent with the tribal falconers of Pakistan in 2007 that overwhelmingly changed my perception of birds of prey. Like all indigenous people, the tribal falconers' methods have remained unchanged for many thousands of years; their form of hunting with hawks is perhaps the purest I have ever experienced. On the whole, they are poor subsistence farmers, falconry forming part of their survival. For these people, falconry is deeply ingrained within an identity, which itself is set in equal balance with their environment.

My time in their company exposed me to a way of life little known to the Western world. Despite the obvious language barriers, seemingly insurmountable cultural differences and customs, we shared, through falconry, an instant kinship, became friends and respected one another.

Some of the lessons I learnt have now been translated into the training and rehabilitation of the hawks in my cottage. The Pakistani falconers' methods are ancient, evolved and so refined that on these hawks' exquisitely formed shoulders rest generations of shared human history. They are perhaps the most important hawks I have ever owned. They are utterly irreplaceable, priceless in a very literal sense, and I cannot wait to set them free.

I hope I never see them again.

Looking back, it seems inevitable that I would be a falconer.

As a child, our household had respect for different races and cultures, and stories of travel pervaded our lives. In his youth, my father followed the hippy trail through Europe and down into India. He travelled to America and also worked for mining companies in the outback of Australia. He kept mementoes of his travels in a huge oak chest. I would climb inside and spend hours touching and looking at these strange, otherworldly objects: fossilized wood, a desiccated puffer fish; a spiky ball of pale tan and tinted yellows; a dingo's canine tooth; coins; beads; wooden totems; and numerous black-and-white photographs. One image remains in my mind to this day. Standing in a desert, two lithe young men hold a pair of protesting eagle chicks – their wings are pulled apart by the men and extend to waist height.

My mother and father's approach to parenting was non-conformist, libertarian and not without turmoil. All rules were fluid, fluctuating on a whim, practical jokes common. They once convinced me I had the power to lay eggs. A nest was built and I was encouraged to sit and wait for one to arrive, like a broody, five-year-old human/hawk hybrid, dressed in a pair of Spiderman underpants. I can still feel the tense curl of expectation in my stomach. The inevitable disappointment was huge.

Our cottage was situated deep within the English country-side. Exposed to nature at an early age, artistic and creative, I lived in my imagination or in the outdoors. I built numerous dens and hideouts, weaving wood and leaf litter, cutting and creating my own world in the process. I made fire, would hunt and trap animals. I taught myself to tickle brook trout, whipping them out of water by hand. Once they had been gutted, I flash-fried them for lunch. In the spring and summer, slowworms, newts, frogs, toads and gallons of tadpoles ended up in buckets. I threw stones at a hornets' nest, and big brutal insects spilled into the sky in a massed ochre cloud, the buzzing tone dangerous and low. I tied daddy-long-legs into cotton thread, flying them high into the sky, before winding them back to the ground, joyfully repeating the process for hours. An injured mole, cleverly kept and watched in a metal box, was fed worms I dug from the garden. Unable to keep up with her voracious appetite, I eventually released her. I clearly remember her soft silver coat and fat, pink-paddle hands as she swam back beneath the soil. Numerous tennis-ball-sized baby rabbits arrived on the floor and doorstep. Some survived, others succumbing to the stress and mess and the cat's jaws. On a low winter afternoon, alone among snow-stripped saplings and bare trees, a muntjac deer screamed. In front of me on the floor, an isolated, trident-shaped footprint. A head full of my own fantastic stories and the strange noise of the deer, I knew for certain it was a monster. I panicked and ran. Overly inquisitive and without an active idea of cause and effect, I dipped my hands into the depths of a fallen tree. I removed a palm-sized baby squirrel; it was cold, almost lifeless. To raise its body temperature, I tucked it into the only thing warm enough, my sweaty plimsoll. I ran the half a mile home barefoot, scuffing and blistering my feet. Later I fed her cow's milk but failed to keep her fragile life from slipping away. From the long grass in our garden, I 'borrowed' twenty grey partridge chicks from the nest of their strutting, calling mother. Each was the shape and texture of a bumble bee, and I built them a new home across my duvet, using a hairdryer to keep them warm. Once discovered, they were boxed up and driven to a member of the local shoot, to be raised, released then shot: the oddest parental logic. Smaller birds were a better prospect, some fallen from nests, others brought in by the cats. I raised one by hand; it would perch indoors on the curtains, flying across the front room for worms and hand-held maggots and was a potent shape of things to come.

As an adult, life has not been straightforward.

I fail to maintain close long-term human friendships, or intimate relationships. I dislike crowds and large groups, much preferring my own company for extended periods of time. Visually, I am hypersensitive, amplifying and magnifying all communication and, with no filter, it often takes days to decompress and unravel meaning. When talking to people, I have multiple interpretations of one conversation. I will avoid eye contact, become distracted and agitated easily, my levels of fear and anxiety are often unbearably high; around strangers, I am in a fight-or-flight position for much of the time. Unconsciously, I will overstep the mark, push boundaries and blurt out inappropriate comments. This often makes me seem impulsive and anarchic.

I also develop relentless sets of routines around subjects that interest me. To waver or break in them causes me distress and frustration. I explore these subjects at the expense of everything else, exhausting myself and anyone else close to me.

I'm a natural outsider, unable to relax yet struggling to meaningfully connect, but nature has continued to intercede for me as a place of peace, a welcoming conduit of stabilizing emotion – a place where I feel most able to express and communicate who I am.

I find nature to be infinitely absorbing and visually relaxing. I am utterly in thrall to the in-built freedoms and multiplicity of the natural world. The mesmerizing fractal nodes and colourful noise, the giddy rush of detail, the delicate points of pattern in the forms of animals, plants, elements, tastes and textures, all make perfect sense. Deeply democratic, all that scuttles and swims, sucks, prowls, bounces or blows, everything that hatches, pushes, pulses, flies, fans or breathes is of interest to me. I am in love with the endless creativity that throws up varied forms, billions of ideas that flip and fold, live and die, survive or pass. The natural world is the embodiment and perfect playground of difference, a force celebrated simply by and for itself, a place without boundaries or fear, out of which I have built a self-willed and life-affirming education. Nature remains the most significant, meaningful and consistent relationship of my life. Without it, I feel helplessly lost.

My discovery of birds of prey was a revelation. When I held a hawk for the first time, shocked by an emotion of such startling power and clarity, I felt an internal, audible crack. This was what I had been searching for.

Over time, the detailed characteristics of the hawks slowly emerged. The powerful connection I experienced made complete sense. Hawks are very pure creatures, highly nervous, intelligent, fearful, and loners for large portions of their lives. They live in the moment, possessing few subtle grey areas, acting or reacting on an in-built hardwired nature. If handled inconsistently and without care, they return to a wild state, quickly. A good relationship between a hawk and a human has definite, intimate parameters, parameters absolutely set by the hawk and not by the human. They do not waver, cannot be bullied, coerced or negotiated with. They possess their own internal logic and have very specific requirements. To succeed in building a strong relationship, all human ego must be subjugated in favour of the hawk's needs. When training a hawk, you have to think through them in order to create a workable bond. You have to surrender and enter their world, understand it through their eyes and serve them with patience and a deeply held empathic equality.

Anything less and you will fail.

It was after watching the falconers of Pakistan, and participating in the lives of other falconers, that the facets of my feelings of identity with nature and my understanding of hawks all fell into place. It was their methods and ways of living that drew everything together. Among these people, in differing degrees, I found an attitude reflecting my own. It was through the merging of East and West, ancient and modern, that a better understanding of birds of prey, their quarry and conservation, began to materialize for me. My experience in Pakistan showed me that falconry is a connected patchwork of experience, is interchangeable and mutually supportive, no matter which culture is involved. And the highest, most durable, most transcendent falconry, regardless of the continent, stems from the same fundamental, ancient source: indigenous wild hawks, flown over wild landscapes, hunting indigenous wild quarry.

Perhaps the most important fact I took from my trip to Pakistan was the simplicity and freedom of their falconry; that their lives revolve utterly around the hawks: everything starts and stops with their immediate location and landscape. To witness humans deploying 5,000-year-old methods with style and humour, taking only what they need and never leaving a trace, was a truly beautiful thing. From this I came to realize my deep human instinct to hunt is morally correct, that hunting with a hawk is not cruel, unusual or destructive but an activity that replicates and exists in harmony with nature: that harvesting food using hawks is a gift and privilege, conducted in a wholly balanced way.

On my return from South Asia I entered a slow transition, unfolding and unpicking the journey into a workable logic. The experience stored itself away like a seed, the soft kernel of buried knowledge waiting to split as time passed. Almost four years later, in 2010, the shattering news arrived of a terrible, violent event. The tribe, the village, their children, the dogs and the hawks, were displaced by unprecedented flooding. A direct effect of climate change, a living, breathing, history of agriculture and falconry was wiped out in a matter of days.

This event prompted a quiet, motivational anger (which remains part of me). This was not some abstract news story – turn the page and forget about it. In my own peculiar way, I elected not to forget and, enriched as I was by the experiences of wider travel, the tribal falconers' methods and approach to falconry eventually transformed my life.

It would be a lie to say this transformation was planned, or in some way a considered lifestyle choice. It was not. The rooted circumstances and triggers were beyond my control, arriving with a sudden drop and from an odd, unexpected angle in the same year the devastating floods arrived in Pakistan.

Viewing my story from a distance, I observe an alarmingly steep emotional descent at the point my son was born. At a moment when most people celebrate and rejoice, I entered an altogether different, darker, more violent, psychological journey, my failures and frustrations underpinned and set in motion by the then unrecognized aspects of my personality. At the time I was not conscious of my descent and was to a large degree powerless to prevent the free fall. But as I fell I committed myself to a course of action that horrifies and angers society. I walked away from my son without any idea of how, when or even if I would ever return.

It would be disingenuous to say that falconry or rehabilitating wild hawks helped me at this time, that, in saving them, I somehow saved myself. That is the stuff of fiction. When the pain and guilt was at its worst I had abandoned not only my son but also birds of prey, returning to both only when I had wrestled myself free of the emotional wreckage my life had become. This freedom was pivotal – it sustained my mental clarity and calm, through which my love of the natural world provided a stable framework to rebuild and regenerate. From this platform I kept climbing, recalibrating and joining the dots. My journeys, the lives of the falconers I met and my very own existence ceased being separate stories and coalesced into a free-wheeling road map, a way of living and a way of understanding myself that I should have started on many years ago. I saw that my feelings towards nature, and birds of prey in particular, ran in parallel with my feelings for my son. I worked out that they were, in fact, two sides of the same coin – the deep love of one could, with gentle observation, inform and unlock the deep love for the other. Out of this realization I tentatively began to re-form a relationship with my son and his mother. I learnt to overcome my fear of fatherhood and found a way to positively express myself through and for my son. Perhaps this then is the central theme of my story.

My life is now defined by the hawks, their moods, the seasons, their landscapes and the quarry they hunt. I have finally found a place apart where I can rehabilitate and release wild sparrowhawks and hunt goshawks in the same manner, and for the same reasons, as found in Pakistan.

I have stripped away everything I consider superfluous. I have turned my back on a well-paid career and a post-graduate teaching certificate from Cambridge University. I live as simply as possible, earning a modest income as an artist and through seasonal summer work on country estates. I live with two dogs. My cottage is down an old bumpy track on the outskirts of a tiny village. I own no substantial property, I have no debts, no credit cards and little in the way of modern amenities apart from an old laptop. I have no double glazing, no central heating, just a large log burner in the front room. When I am cold, I collect wood and make a fire. I use little electricity, have no television, no phone landline, no Wifi, no internet and no distractions. When I am hungry I walk into the fields and go hawking. If I fail to kill, I try to trap other animals or catch fish. I have fruit trees and three small vegetable patches. I own a freezer and a fridge and buy other vegetables seasonally from farmers' markets. If I require anything else, the village shop and post office is three miles away, the nearest supermarket six.

This is not a life that would suit everyone. I feel the elements and the seasons directly: autumn and winter are cold, dark and long. But when spring arrives, songbirds nest in the roof and produce offspring so numerous and so vocal the whole cottage literally sings. On summer evenings I sit on my front step and watch four species of indigenous bat flit and skip across the dusk. In a half-mile radius, there are foxes, frogs, natterjack toads, great-crested newts, grass snakes, wild sparrowhawks and nesting goshawks. Flies fly in through the door and I watch the spiders in the corners catch them. I have rats in the roof, a hedgehog that hibernates behind the chimney. Nearby, there are peregrine falcons, marsh harriers, red kite, buzzards, kestrels and merlins. Stoat and weasels regularly streak past, jerking and hopping, hypnotic when hunting. The fields and coppices contain pheasant, rabbit, curlew, lapwing and lark. Hares box one another in March. The brook is awash with insects, spawning brown trout, salmon and a dozen species of duck. Kingfishers nest in the bank running along the stream. Elvers from the Sargasso Sea, translucent silver shards, swarm up the dam to the many silt-laden ponds dotted around the cottage. Indigenous crayfish, little armoured fresh-water lobsters, hide tight, tucked under rocks in the nearby reservoir. Both native species of wood-pecker zip past with a looping, swaying flight. Bullfinch, collared dove, greenfinch and dozens of hedge sparrows bob about in the garden. Because the cottage is situated on a mig-ratory path, thousands of greylag, Canada and pink-footed geese fly honking low and loud over the roof. Here the natu-ral world is abundant, bursting, squabbling, chirruping and stretching at the seams, inside and out, all year round.

A life lived through and for a hawk is ever evolving, ever surprising, a sensuous experience based on rhythms in tune with natural cycles. It is a life awake with potential, always exhilarating. A life that readily sends me into the landscape, into nature, to think, to feel and to wonder.

1

The Road to Pakistan

A bell for a hawk is small and significant.

A good hawk will kill at tremendous distances, feed in thick cover, silent, camouflaged and almost invisible. When the falconer finally arrives, the high-pitched cadence of a bell is often the last lifeline between location and loss.

In the distant past a falconer would make all their own equipment. Gloves, hawking bags, lures, swivels, perches, hoods and most certainly bells, all bespoke, all unique, all tailor-made to suit. Now mass-produced and machined, convenience has put paid to the individuality of handcrafted bells. Among falconers, there is great prestige in making your own equipment, particularly if the process is a lost art. Naturally, I became obsessed with rediscovering how to make bells and quickly ordered all the equipment without any real knowledge of how to construct them.

For over a week I worked wildly, using intuition and learning from my mistakes. I coated the house in a fine metallic dust, razor-sharp strips of metal became trapped in the floor, cutting feet. I burnt holes in the carpet. Too focused and frenzied to bother with safety equipment, chunks of metal scattered like stars, flew from the Dremel and embedded in my forehead, creating a tattoo of dark metallic spots beneath a red-raw rash. Some bells looked beautiful and worked; others fell apart almost immediately. To refine them and find consistency, I went to the library for books and turned to social media for help.

The Arabian and Muslim nations have a long and intimate connection with falconry. Birds of prey spiral through the very core of their way of life. Long before the West knew the potential of hawks or falcons for hunting, the Muslim world was perfecting and turning the practice into high science. They helped introduce the practice to Europe through trade routes and the prophet Muhammad is said to have been an avid falconer. It is no surprise, then, that the oldest styles and most traditional falconry bells are still handcrafted in the East.

After searching falconry forums, I found images of intricately carved and jewelled bells, the artistry and design detailed and beautiful beyond belief. The gentleman selling them was a falconer from Pakistan. So I emailed the seller, a man named

Salman Ali. Over the course of our correspondence, I asked to come and see how they were made. He agreed so, in 2007, I emptied my bank account and booked a flight. It was as impulsive and as simple as that.

I estimated the trip within Pakistan would take two or three days. I would check into a hotel and be there and back within a week. Outlining the small dangers of Karachi, Salman instead invited me to stay at his house, remain in Pakistan for a longer period and fly goshawks in wilder parts of Sindh province. We would then travel to Lahore and meet the bell-maker. I had no real idea of the consequences this generous offer would bring. I simply took a Muslim stranger at face value, my well-being now in his hands.

A few months later, a quiet, smartly dressed, muscular man warily shook my hand at the airport and we drove to his house in the affluent suburbs. Heavily disorientated and jet-lagged, I crashed out into a long sleep. The next morning, in a small, tidy courtyard under the shade of a large fig tree, we sat sipping tea.

In the walled yard, Salman had established a modest falcon rehabilitation centre. Over his shoulder, four wild falcons were blocked out on perches: a saker, a smaller male of the same species; a sakret; a lugger falcon; and a species of peregrine with the beautiful name of red-naped shaheen (pronounced shaar-heen).

These species of falcon are highly prized on the Arabian

Peninsula, some individuals fetching up to £35,000 or more. Over the centuries, sakers and luggers have been trapped in large numbers, transported, exported and sold illegally in animal markets, or to rich, private individuals in the Gulf states. Many die in the process and, coupled with increasing habitat loss, all of them are very much under threat.

Once healthy, these particular falcons were to be re-released and would potentially go on to live long and productive lives when released. Unfortunately, it was also likely they would once again be illegally trapped, finding themselves back in the market place. Conservation uppermost in his mind, Salman reasoned the slight chance of success was well worth the time and effort.

A healthy falcon of any species has a formidable and impressive presence, their eyes clear dark orbs, their feet clean, powerful, scaly, lizard-like tridents, capable of snatching prey at speed in mid-air. The beak should be smooth, clipped, tip sharp and curved. The shape and appearance of a healthy falcon is akin to warm air rolling over a summer crop. From head to tail, as you follow their outline, feathered contours flow with gentle indentations, smooth undulations; no breaks, no splintered edges. A fit falcon is a feather-perfect inverted teardrop.

The falcons in Salman's yard were misshapen, scummy, a dull colour. They emanated an inherently odd impression, dissolved at the wrong angles, stood badly, eyes tired, cloudy;

they seemed almost half built, restrained and infinitely sad. On inspection, they had various ailments: broken feathers; fractured and split beaks; twisted, cankered, pox-riddled, scabbed and sore feet. Distressing to see, these external injuries were only the tip of the iceberg.

Hawks and falcons rarely show weakness unless close to death, a natural safety mechanism preventing predation in the wild by other animals. It requires keen observation and hard-won experience to ascertain and understand the root causes of the deeper diseases lurking beyond the obvious.

To be rehabilitated correctly, a bird of prey requires a consistent supply of food. It can take many months to nurse a hawk or falcon back from illness, and they must be fed daily. England has specific producers of raptor food, much of it a by-product from our egg-laying industry. All the protein is clean, bacteria free, cheap and delivered to the door at the click of a button. Once defrosted, the meat can be fed directly, without any fear or worry of cross-contamination.

Salman was using feral street pigeons.

Running short, we went to purchase some from the animal market. Down a honeycomb of dark, medieval passages, hundreds of square mesh cages were stacked column by column, twenty feet high. Monkeys, birds, falcons, rodents, owls, lizards, all pushed and packed inches apart, the noise, smell and heat cloying. Piss and shit dripped and splattered down through the wire, on to every animal, top to bottom. I felt something tugging at the edge of my trousers. Fat rats slipped and bumped across my boots.

The purchase complete, we hauled the plastic sack of wriggling and restless birds to the jeep. We spent the rest of the afternoon butchering them by hand, inspecting each for disease. One scrawny carcass contained a large yellow sack of rotting fluid around the heart. Salman threw it away. In its place, we killed several wild trapped sparrows which were bobbing about in cages behind the house. Once fed, the falcons relaxed, and we applied prepared acacia paste (a natural antibiotic) to their feet and left it to do its work.

At midnight, I was told to pack a small bag, and we set off to the coach station.

Outside the city limits a dozen hours pass as we travel in blackness. As dawn breaks, Sindh province emerges disconcertingly, large fields of kilns belching fire and smoke, burning and turning clay into solid workable bricks. It looks like a wasteland.

At the last coach stop we take a taxi, arriving at the side of a road near a tiny village. The landscape has changed, now fecund, stretching flat, a vast strip from left to right. Irrigation ditches streak horizontally to the distance, trees and shrubs are dotted near and far. In front of us is a low nondescript wall, in its centre, partially hidden by a palm tree, a door.

Inside, a large rectangular compound opens up. Three modest homes are haphazardly built into the corners. Along the wall two stables, a donkey, a latrine. A dozen tough-looking chickens and a bitch with puppies stroll and scamper about in the sun.

Four men – Manzoor, Chanesar, Jamal and Haider – greet us. Their children bring two raffia rope beds out to the centre of the compound and we sit. Three of the men disappear into their houses, re-emerging moments later, each carrying a large goshawk.

The hawks look about, searching for imperceptible movement, blinking at the now bright world around them. Two of them, a male and a female, have tinged red orange smeared across domed yellow irises, a gradation of colour as subtle as ink separated on blotting paper. Their feathers are a mix of light browns, with touches of slate grey across the wings and back. These darker, bluish feathers pop up sporadically in patches, following no particular pattern. The varied coloration indicates they are passing into maturity, their adult feathers slowly emerging, becoming more prevalent, resplendent.

The smaller female has a pure plumage of caramel cream, light brown and touches of white. She is muted, camouflaged in comparison with the male. Fully grown, but hatched less than eight months ago, she is a juvenile of that year.

Extending from the side of each goshawk are short feathery trousers, that look like culottes. From the base of the culottes, legs and feet extend thick, dense and powerfully strong. The central toe, measuring roughly three inches, is as wide as a man's finger. A keratin-black talon, curved and scythe-like, doubles the length. Under their pale, yellow-scaled skin, a mass of sinew connects the toe and talon to muscles at the top of their legs.

These goshawks have been trapped in the wild. They are the first wild goshawks I have ever seen and, although trained, they teeter on a precipice between domestication and un-adulterated instinct. Prior to their contact with humans, they had grown and evolved out of the land. At rest, they present a paradox, equal parts beauty and destruction. They have a compact, raw power underpinned and overlaid by delicate feathered fragility, a detailed, subtle softness unexpected in something so deadly. They smother the space with their presence, their outward calm clearly indicating they are at ease with the surroundings. I find them hypnotic.

Before this moment I had seen many goshawks, all man-made, using artificial insemination or bred in captivity. In comparison, these hawks radiate difference, each possessing a quality uniquely original, something liberated, a life force well beyond human rules and control. They are a celebration of total freedom. Their hard-edged vitality, their context and their history shifts my world on its axis.

To understand why, you need to know the roots and history of falconry.

Over 5,000 years ago, on the cusp of the very first human kingdoms, prior to the first written words, before textual history and long before the concept of pressed coins and the idea of paper money, well before organized monotheistic religion, 2,600 years before Islam, and 2,000 years before Christianity, a human or humans on the Russian steppes watched a bird of prey make a kill in the wild.

Walking down the valley towards the raptor, they startled the hawk and it flew away, leaving the quarry behind. The freshly killed carcass lay half eaten on the ground. They picked it up, looked at it, took it to their camp and ate it. They survived the night. Not content with finding haphazard kills, these humans decided to trap a wild hawk, train it and use the hawk's natural hunting abilities to provide food for their families.

Falconry was born.

There is no written account of this ever happening, but images of humans hunting with birds of prey exist from this time. Along with dogs and horses, hunting with birds of prey represents one of the oldest examples of humankind working with animals in a mutually supportive relationship. The three goshawks in front of me are directly related to this unbroken lineage. The processes by which the tribal falconers trap, train and hunt with these hawks is the same as those used by the first falconers on the Russian steppes. These goshawks are an essential time capsule, underscored with information that has crossed continents and cultures. Their training and existence are a handshake from ancestors unknown, a near-ninety-generation meme surviving the test of time; still living, still a reality, still vital. I am humbled to be in their presence.

I watch as Haider's smaller male goshawk notices the chicks pecking at my feet in the dust. The hawk, following their movement, turns his head in one slow, fluid motion. A change of mood washes over him: his feathers noticeably tighten, then relax; he rouses, then yawns. It is clear he will soon be ready for hunting.

A few hours later the sun has dropped to a bright white disc and with it comes a slight drop in temperature. The timing and change is important to the hawks. Haider rises and moves in a direct, less cordial manner. Chanesar, crouching, examines the floor, studying streaks of hawk faeces in the dust. Normally a black solid in a white liquid surround, it has changed to viridian green foam on an opaque chalky background. This mutation in tone and texture indicates the hawks' digestive tract is empty of yesterday's food.

A lack of protein and the time of day has triggered the hawks' appetite. They are visibly restless: the feathers on their heads rise to a low crest, the edges thin, jagged like a the outline of a dried teasel. The hawks' heads begin bobbing, jerking; their pupils expand and dilate. They begin to enter a very specific behavioural phase called *yarak.*

In the wild, if a hawk exists on a constant knife edge between starvation and success, if it lives solely from one kill to the next, it will be weak, inefficient and die. When hunting naturally, a hawk fails more than it succeeds. Nine out of ten attempts at securing a kill end in failure. A hawk therefore needs stamina as much as speed, it needs to be mentally alert, focused and unrelenting when hunting. Any wild hawk surviving to maturity has copious fat reserves, is heavy-set and in perfect condition. Hunger for a wild hawk is not necessarily a motivating factor; they hunt because it is their sole meaning and purpose, for fun, even. The desire to fly and kill without the twist of hunger, to fly on appetite alone, is the mark of the most successful, most powerful and healthiest of hawks.

Achieving *yarak* in a falconry hawk is therefore highly technical. Only the most skilful falconers attain it with any level of consistency. A heavy, overweight parent-reared or wild trapped hawk will not want to be near a human and will fly off. To achieve *yarak* takes time and dedication. The hawk needs to be conditioned correctly, flown daily and to make a kill at least once every other day. To succeed, the hawk and falconer require time, space, land and an abundance of quarry.

I watch as Haider picks up his goshawk and holds it close

to his face. Looking intently into the hawk's eyes, he pushes his fingers deep through the feathers on to the chest. The goshawk makes no move or protest, is completely relaxed. Haider is feeling his weight, watching his reactions, through observation and touch, intimate indicators emanating from the hawk. The hawk is telling him in detail how deep in *yarak* he is, and so how close to hunting.

Later, I feel the hawk myself. He is barrel shaped. There is no bony edge to the chest, just a thin strip of keel; the muscles are compact and round, extreme with fat. I have never felt a hawk as fit and well prepared for hunting, before or since.

With mounting activity, the tension in the compound is palpable. The extended tribe arrives, milling about near the entrance. Our hawking party is growing, including friends, family, children and elders. Having spent most of the day walking from a village several miles away, dressed in a neat shark-tooth jacket and sporting a fine fake TAG-Heuer watch, a falconer named Punhal arrives with his female sparrowhawk.

A sparrowhawk is more or less a goshawk in miniature. They are both of the *Accipiter* family, with similar psychology and behaviour. The only significant differences are that the sparrowhawk is explosively faster across shorter distances, and the quarry it kills. The goshawk has evolved to kill larger birds, rabbits, hare and more or less anything that hops, flies or runs and weighs under three or four pounds. The sparrowhawk, on the other hand, is adept in and perfect at thinning out smaller birds and mammals.

With all the commotion, jokes, chatter and general movement, the hawking is turning out to be a significant, connected, communal activity. I expected the numerous people to simply wave us on our way with the hawks. Instead, the whole group is intending to head out and assist, the energy of our party rising from a shared purpose, the default setting of hunter-gatherers the world over.

As we load the jeep the hawks are swapped between the different members of the tribe. Despite having primary handlers, the goshawks and sparrowhawk seem to belong to the whole community. The water-carriers and beaters appear to be as important as the falconers themselves, each person an integral part of the hunting group. In all, I count thirteen people, three goshawks, a sparrowhawk and two shotguns crammed into two vehicles.

When I ask why they share the hawks the answer given is as logical as it is simple. If the hawk responds to only one falconer, it may be lost or killed if that individual owner cannot reach it in time. If a specific hawk belongs to a particular falconer and that falconer falls ill or breaks a leg, then the hawk cannot be flown by anyone else; it becomes next to useless. Better to have thirteen people who can handle a hawk equally than just one.

The jeeps pull out of the compound, the village drops

away and the landscape loses evidence of overt human interference, becoming wilder. Twenty minutes later we stop. The
falconers and beaters climb from the vehicles and stand in
the road surveying the landscape. Salman nods and tells me
the hawks are ready.

The group splits in two, and the big female goshawk leaves
with Salman and Chanesar. I remain with Haider, his male
goshawk and three other assistants. We move off the road,
down into rough terrain, the vegetation chest high, soft and
silky, dense, a radiant green.

This area is the habitat of a small gamebird, highly prized
by the falconers. The francolin, also known as the black partridge, is the size and shape of the indigenous grey partridge
found in England. It is tough and flies with speed and agility.
A francolin is difficult and complex to catch, and the equal of
any goshawk.

Our group spreads out and we begin walking slowly.

Haider holds the goshawk high above his head. Almost
immediately, the bird becomes agitated and changes his
behaviour. He stretches up, thin on long legs, his head jerking
back and forth. His neck extends far forward, his feathers
contracting tight. Flapping violently, he whips back around
on the end of leather straps (jesses) attached to his ankles
and held in Haider's gloved hand. He has seen something we
have not and focuses tight on the target. I am close enough
to see his eyes expand, thin lines of pulsing yellow snapping

back against black. His pupils are now the size of a thumbnail, exceptionally wide, able to suck up all light and movement in front of us.

The cover ahead thins. From across the gap, flying hard and fast, a blurred and whirring francolin shears past. The hawk is released, his acceleration instant. Over the first few feet the francolin climbs high, a mounting arc, swift and perfect. The goshawk follows, reconsiders, drops below and slightly behind the francolin. The sun, reflected, briefly flashes across the bloom of the hawk's back.

For the first hundred yards the goshawk repeatedly turns his head skywards. The francolin will either land abruptly or power to the horizon. By holding back, the goshawk is wasting little energy, making no mistakes, watching, waiting, clever and cunning.

At two, then three hundred yards, they remain together, the francolin refusing to roll or change direction. Her pace is metronomic, precise and controlled.

Seconds tick by as the flight unfolds.

At five hundred yards, tiring, fearful, the francolin attempts to rise higher, to outfly the hawk. In shadowed synchronicity, the goshawk follows. They keep moving, are evenly matched. Imperceptibly, the balance of power changes. The francolin panics, the goshawk, slipping fast forward at full tilt, cuts up the sky in silence and intersects the gamebird. As a single black dot, together they drop in slow, stalled momentum.

Sprinting through the heat and heavy vegetation, I reach the area where the goshawk and francolin have fallen. I hear a bell and find the hawk on a small patch of clear dirt between fronds of pampas grass. I watch closely as the goshawk pulls savagely at the francolin's primary flight feathers on each wing. The francolin flaps to escape, the goshawk clamps down on her back. By removing the outer flight feathers first, the francolin is incapacitated. Even if she wriggles free, she cannot fly properly and would be easily recaptured. This is the evolved behaviour and actions of a successful wild hawk, one motivated by pure instinct.

A few seconds later Haider arrives and kneels next to me. Both of us are breathing heavily. We look at each other, smile and laugh out loud. Both of us in equal awe and connected by the hawk's success. Using a small knife, Haider cuts cleanly through the francolin's neck. The blood flows slowly, a thick, deep, oxygenated red. The bird flutters, unwinds like an elasticated toy propeller, a thrumming drum, then lies still. Haider allows the goshawk to break into the chest cavity; he rips at flesh with brutality, then fragile precision. While his hawk feeds Haider gently lifts his tail, preventing the long feathers from bending or breaking on the ground. The goshawk remains at ease, unconcerned by the intrusion.

After a wild hawk has killed and fed, it will find a place of safety to relax. Having eaten his reward, Haider's goshawk looks about, searching out the highest perch. With no trees nearby, a six-foot white western male seems the best place to digest a meal. From the ground, the goshawk launches skywards, landing on my shoulder. Not content, he moves higher still, up on to my head. I feel the sharp-tipped pressure of talons beneath my hat, but no sense, sign or display of aggression. As if on cue, the rest of our party arrive, see the hawk on my head and start clapping and cheering.

The francolin is lifted off the ground and inspected. It is a complex array of browns, dust and mud colours, with flecks of yellow in delicate speckled patterns. Although lifeless, she looks quite mesmerizing. Matched evenly by a creature that evolved alongside her, her capture and death are certainly sad, but also arbitrary. The flight could have gone one way or the other. Today, the hawk has been lucky. What I have witnessed is a truly natural selection.

As Haider and our group prepare to resume hunting, a call goes up in the distance. Salman's party have flushed several partridge. Selecting one, Chanesar's goshawk is moving swiftly, flying hard across the horizon, wings pumping. The francolin, with a better head start, confidently flicks over, dropping into thick, heavy cover. The falconers, running, reach the hawk several seconds later. She is picked up and prepared for the re-flush.

Where the francolin has landed looks difficult to enter. Aiming to scare the partridge into flight, the falconers simply set light to the undergrowth. The flames rapidly take hold and the fire rolls forward; a sparking, powerful, energetic noise. Squealing steam, sap vapour, erupts from thin branches, smoke begins to obscure the skyline. Through the cracking, blazing heat, a large rotating flake of ash transforms into a francolin, swept up and flying fast. The goshawk launches rapidly off the fist, follows the partridge up through the smoke and out across the shimmering landscape. They disappear behind the flames and fumes, the falconers running to catch up. Later, I would be told the flight was a success, the francolin killed nearly a quarter of a mile away. A massive distance for any hawk.

Over the next few hours the different groups flush many more francolin. All escape, and the goshawks fail. We change tack and drive to areas of farming and cultivation, where irrigation ditches have been dug to control the flow of the Indus river's tributaries and direct water to crops. Hanging gently over the shallow canals, puffy green bushes dot the banks. On sight, ducks, water-rails, egrets, herons, bitterns and a coot-like bird take to the wing, skitter across the surface, hiding under cover. They are clumsy in flight, and the goshawks pursue them relentlessly into the reeds, bushes and bankside vegetation. Everything we flush is outflown, the size and power of the goshawks unsparingly consistent. Flight after flight is taken, each ending in a kill. The hawks are not rewarded with blood or fed from flesh; simply plucking at the feathers seems to keep them motivated.

One particular bird lifts up out of a ditch and loops across the soil, using the underside of the jeep as protection. The goshawk simply folds her wings and hits it with force, rolling her prey through dust out the other side.

Each falconer is allowed their turn. Without exception, they take great pleasure in killing a grackle-like bird about the size of a magpie. This species has a long, henna-coloured tail; the falconers have nicknamed it the 'imam bird'. The whole hunting party cheers with delight when one is killed.

Despite the mounting numbers, the treatment of each dead bird is respectful and without sentimentality; a brief pause to consider, like picking an especially beautiful apple from a tree. As time passes, the flights never cease or thin. For every bird caught, three, four or five escape. The hawks, although efficient, can take only one at a time. There are spaces and gaps of rest. When pleased with a stylish kill, we move to a new location. We take no more than we need, and the chance of endangering wild populations is negligible. Watching the beaters and falconers, there is no sense that survival need be a struggle. The energy is celebratory. There are numerous overlapping conversations; gossip abounds, jokes fly back and forth. They are happy, relaxed, comfortable and excited.

By the time darkness descends we have caught enough. It has been an astonishing display. As we climb into the jeep the goshawks are fed their rewards and rested on gloves.

Small boys and beaters hold birds of numerous colours and varieties, poking and pulling at the feet and wings, inspecting and learning.

On the way back we stop at the village to drink tea. The dead francolin and other quarry are sent ahead. An hour later, in a small brick room, we sit with a tribal elder. As I admire and handle a new Kalashnikov rifle small bowls of cooked and lightly spiced meat are placed before us, an incredible feast of flesh. A dark clutter of bones stripped bare becomes a high pile in minutes.

Back at the compound, away from the formality surrounding the elder, the atmosphere is more egalitarian. An electric bulb weakly emits a soft egg of light; darkness cloaks all but close contact. Cigarettes are swapped, and we relax, the happy release of an intense shared hunting experience. The day is replayed through talk and translation. There is intimate laughter, discussions and humorous anecdotes. One is about a troop of monkeys that stole a local man's possessions while he rested in the afternoon sun. He spiked their drinking water with herbs similar to cannabis and they quickly fell asleep, dropping his shoes, watch and blanket out of the palm tree.

Chanesar's goshawk is brought in and placed on a long pole, high above our heads. She recently killed one of the compound's cats. Huge and belligerent, she cannot be trusted to leave the other one alive. A mouse pops out from a hole in the wall and runs across the shelf beneath her. With only one cat, the rodent population is growing rapidly.

Salman takes out the present I bought for him in return for his hospitality. Telemetry is a lightweight tracking device attached to a bird of prey. It will send out an electronic signal from as far away as twenty miles, which means any lost or errant hawk can be tracked down. I had brought the best telemetry set available, all high-tech wires, aerials, beeping, red LEDs and plastic and metal casings. Salman explains to the group how it works and what it is for. There is a slight pause, a brief silence, Haider makes a comment, then laughter. Salman translates: 'English falconers should try to train their hawks better, and not lose so many.' Through this gentle mocking, I feel involved: in the intimacy of this dark room there are no real differences between us, I am just another falconer welcomed in. More importantly, Haider is right; perhaps we should train our hawks better.

Eventually growing tired and in need of rest, I lie back. The remaining goshawks are moved indoors and placed on the edge of our beds. They will sleep next to us until dawn.

Early in the morning Chanesar comes out of his house holding a small bamboo cage containing a male francolin used for trapping goshawks and sparrowhawks. I sit, sipping tea, staring at the caged bird. He is beautiful: jet black beneath a foil of rainbow shades, like petrol spilled on tarmac. Briefly, I look up and out across the compound. A white blob emerges on the horizon. Ten minutes later it morphs into a small boy and his father; a further five minutes and they are standing before me. On the little boy's bare fist is a small, freshly trapped male sparrowhawk: known as a musket. At only 5 ounces, and only 8 inches tall, he is fragile, tiny and roughly a third smaller than Punhal's female.

Of all the hawk species, the musket is the most delicate and difficult bird of prey to train. In the wild, male sparrowhawks are rarely seen, such is their speed, agility and success in camouflage. Their size makes them highly fearful and secretive, and it takes remarkable skill and knowledge to keep one in view, let alone get close enough to trap one. According to Salman, the child and his father caught the little sparrowhawk at dawn, several miles away. The boy is no more than eight years old. I watch the father and son interact, their tenderness towards one another exchanged through the shared care of the little hawk. It is a beautiful moment.

Being hysterical, highly strung and indignant, the musket would have acted with tremendous rage at the trap site. To transport the hawk safely, to stop it becoming agitated, flapping and smashing its feathers, the boy and his father have used a process called sealing to carry it back to the compound.

Untangled from the trap, the eyelids of the hawk were pinched up off the surface of the eyeball. A thorn or needle was delicately pushed through the paper-thin skin, creating a small hole. A thread was then passed through the gap. The cotton was pulled, wrapped up over the hawk's head and tied in a firm loop, closing the eyelids. The musket is literally sealed away from daylight. To my untrained eyes, it looks barbaric, primitive and indescribably cruel. Silently appalled and horrified, I keep quiet, taking photographs, as a crowd gathers.

On my return from Pakistan I asked a highly respected avian vet and falconer about the process. His reply shamed me, deflating my self-righteous posturing at the time. There are no nerve endings in the eyelids of birds of prey. The little musket felt next to nothing. Far from being hateful and cruel, the method of sealing has been learned through precise observation and thoughtful understanding of the physiognomy of a hawk. It is designed to briefly protect, not to hurt, the hawk. The same is true for the process of trapping.

On the surface, to trap a creature whose very nature is flight and freedom seems a deliberately unnatural act. In truth, the action of trapping is brief, forming but one small part of a flowing whole, a tiny stage at the centre of a larger, holistic approach to the movement of life in the landscape of Sindh.

Sparrowhawks and goshawks produce between three and six eggs each year. Once hatched, 50 per cent of the chicks will perish within the first twelve months. Broken feathers, illness, injury, predation and poor hunting skills are very real issues, leading to starvation and death. The goshawks in the compound were trapped in the early autumn and only when fully grown. By this time they have left the nest and are beginning to fend for themselves. Trapping is therefore timed to occur long before the forces of natural selection have taken their toll. Taking a hawk at this point is a lifeline, one guaranteeing protection throughout the whole year. The hawk is shielded from death by daily rations and the shelter provided by the tribe. When the hawk is finally released, the company of humans has given it a better head start for survival; for finding a mate, breeding and thus furthering the population of wild hawks. Trapping in this manner is a measured conservation activity, leading to greater life expectancy and therefore freedom for both the hawk and the human.

I watch the boy and his father walk the little musket across the compound into a darkened room to begin the process of training. The crowd slowly disperses and further details about trapping are discussed.

A trapper from any culture needs three basic pieces of equipment: a noose, a net and bait. When using a noose, the process is deceptively simple. The trapper attaches hundreds of two- or three-inch nooses to a domed cage, a bal-chatri (an Indian word meaning 'upturned umbrella'). Inside the bal-chatri is a live bird. The trap and bait are placed in a suitable location. The hawk, seeing the prey, swoops down to kill it. Safely contained in the cage, the bird is prevented from being harmed. Unable to grab the bird, the hawk repeatedly kicks out in frustration, trying to foot the free meal. The hawk's toes become entangled in the small lassos and, as the hawk flaps its wings, attempting to escape, its legs are pulled from under the tail. The gently curved dome of the trap prevents snapped bones or broken feathers and keeps the hawk safe until the trapper arrives.

When using a net, or dho-gaza, the process is equally simple and effective. The nets are made with gossamer-thin threads, almost invisible to the naked eye. The nets are large; roughly eight feet across, and either square or rectangular in shape. Two wooden poles are sunk into the ground with a thin cord strung between them like a washing line. The net is then stretched between the poles, like a vertical trampoline, and lightly clipped to the cord and poles. A bait bird is pegged to the ground behind two or three sets of dho-gaza. Spying an easy meal and unable to see the nets, the hawk or falcon swoops to capture the flapping bird. When it hits the nets, popping them free from the light clips, the hawk is entangled and comes tumbling to the ground.

Theoretically, trapping any bird of prey is as simple as that. The practice is far more complex.

It is not enough to place a trap in a field and hope for success. What separates theory from practice is the knowledge of how a hawk behaves in the context of its own biology, and within specific ecosystems. Trapping is selective, a considered and passionate culmination of lengthy experience, of deeply held human curiosity and a total understanding of the local environment. A good trapper will know nest sites and feeding grounds. They will have built up a large and complex picture of how the hawk or falcon functions. A trapper must possess a complete knowledge of lifecycles, moods and movements, and places of safety. They will know where hawks bathe and roost and at what time, and from what position they hunt, pluck and consume prey. If trapped too early, or too young, the hawk may be noisy, aggressive and difficult to handle. It may not have learnt to hunt with skill and will have difficulty killing suitable quarry. Even if the trapper gets it right, the hawk or falcon may be ill, or have an unseen injury. If so, it is released immediately and the whole process starts again. On top of this, every species of raptor, from an eagle to a sparrowhawk, behaves in a different way. Even within a specific clutch, all siblings vary enormously in their reactions to the world. All trapping practices are specific to the type of hawk or falcon trapped, the landscape and the natural fauna the hawks prey on. Trapping requires sensitive observation and inquisitiveness and is the fruitful product of its own evolution. It changes subtly from culture to culture. Like a musical instrument, trapping has different tones and textures, depending on who chooses to play. All methods are equal, all transformative, with the end result the same: a hawk or falcon on a gloved hand; the purest of notes played. At its best, trapping is advanced field biology, developed and perfected long before this area of science was formalized.

As one would expect, knowledge of this calibre takes many years to accumulate. In Pakistan it survives by word of mouth, from generation to generation. A loving father walking his eight-year-old son through the dawn is just the beginning.

The details of the trapping sites the tribal falconers use are closely guarded information. To trap a hawk can take a long time, sometimes weeks. Despite my obvious enthusiasm for trying to trap a hawk, Salman explains it is not the right time. I had to wait until my return from Pakistan in order to experience trapping first-hand. But what was being explained to me in the compound corresponded exactly to the experience I later had when trapping in America. Despite being on a different continent, in a different culture, we used methods replicating the oldest traditions of Pakistan: a perfect example of the overlapping interconnectedness of falconry. More importantly, the experience I had in America outlined the long tail of history and hidden influence Muslim culture has had on the West.

The falcon I trapped was caught near the tribal prairie land of the Lower Brule, the territory of the Sioux Indian nation. I used a variation of the bal-chatri, a device called a pigeon harness. A tiny leather rucksack covered in nooses; a long nylon cord and a heavy weight coil from the back. The rucksack is strapped to a live pigeon, which attracts the falcon.

On the attack, the falcon catches the prey easily, entangling its feet in exactly the same manner as in a bal-chatri, the heavy weight preventing the falcon flying off.

A South Dakota winter is harsh and cold, wind whipping the temperatures down to around minus 30°C. The prairie grass and all other vegetation is brittle, frozen and coated in a fine, zinc-white snow. It took two days, driving carefully, in white-out conditions, to reach the trapping grounds.

In the warmth of the deluxe four-by-four, and in preparation for trapping, the harness was slipped over the pigeon. Driving the long, straight roads, we looked for signs of fresh kills or falcons in flight. Several were spotted in quick succession, but they were either too old or had injured feet from hunting wild prairie dogs and chipmunks, so were left alone.

Towards late afternoon on the second day a prime-looking juvenile was spotted perched on a T-shaped electricity pylon. From a safe distance, and as seen through binoculars, she had no malformations or injury; all was clear. Eight hundred yards beyond her position, the truck was turned, the straps re-checked, and the pigeon folded up tightly. Slowly passing beneath the falcon, the pigeon was tossed from the window. Once we had left the immediate area, the pigeon began to flutter and pull at the end of the cord. The sight of this sudden meal was too much for the falcon to ignore. From her perch she dropped down and smoothly landed on the pigeon, slicing through the neck, killing it cleanly. We let her feed.

Thirty seconds later the falcon jumped up and lifted off the ground, dropping the pigeon. Free from the trap, untroubled, she circled back around, perching several pylons away. Leaving the truck and examining the lifeless pigeon, I saw that a large chunk of breast meat had been removed with surgical precision, the snow stained a deep luminescent red. The conditions were hard and she was clearly hungry.

Having only one pigeon, now half eaten and dead, the harness was of no use; improvisation and quick thinking were required. The bloody, warm carcass was removed from the harness and tied beneath a two-foot square of chicken wire, the nooses on the upper side, a bal-chatri unfolded and flattened over her kill. Leaving the trap on the ground, with the carcass in full view, we moved away. Five minutes passed and the falcon took flight, dropped down, pumping hard into the wind, and landed on the trap.

Visibly entangled, in fear, fury and with powerful strokes, she rose up with the dead pigeon and trap into the sky, hit a gust of wind and glided low out over the prairie. In our haste the trap had not been secured to the ground correctly. If lost, the falcon would certainly die a painful, protracted death. In the distance, she lifted again and flew another three hundred yards, landed, lifted again, and flew another significant distance. The situation was becoming dangerous and we were losing control.

Not content to let her die, I jumped from the truck and

started running, the mess of moving trap and falcon now bouncing across the frozen grasslands three to five hundred yards away. At the sight of a frenzied human running, she lifted again and made another powerful flight. Momentarily blinded by the glare of sun on snow, when I re-focused the falcon had evaporated into the landscape.

It is a misconception that a prairie is entirely flat. Similar to a ploughed field viewed from ground level, the deep troughs of a prairie are hidden from view. Such is the vast breadth and width of the landscape, it creates an optical illusion of supposed flatness. In reality, this prairie is covered in furrows and dips deeper than a London bus. The falcon could have been a hundred yards away, distressed and injured, and I would not have known. I stopped to listen. As time ticked by, I felt bereft, and I felt anger. If she died, lost and alone, it would be my fault.

I waited.

Usually when a hawk or falcon goes to ground, crows and rooks will begin circling and mobbing; a natural reaction to an unwanted bird of prey. High in the sky, three dots appeared at different corners of the compass: north, east and west. Danger and predation were arriving at an alarming rate. Looking to the rough area where the dots would collide, I started running. On the cusp of a slight incline, the falcon was spinning about, tied to the trap. As I reached her a wild peregrine, a ferruginous hawk and a golden eagle were skimming in towards the stricken falcon. I shouted and attempted to wave them away, the three-tier circle of raptors wound around in the sky, no more than thirty feet above my head. Instinctively trying to protect the falcon, I pulled her and the trap tight to my chest. As I turned my gaze skywards, the eagle flew in dangerously close and low, the white shine of snow reflected across the moisture of its eye. The hawk, eagle and peregrine, unprepared and unwilling to deal with a human, wheeled off in different directions, in search of an easier meal.

In my hands, the warmth of the falcon's body was a sensation beyond words. I untangled her from the trap and held her at eye-level. Common to all prairie falcons, she had a large, streamlined, flat head and a pronounced bony ridge above her eyes. Her beak was strong and light with a pale blue-grey colour. She opened it in protest, revealing soft ridges of flesh and a tongue shaped like the tip of an arrow. Her breath, hot and sweet, smelt like a carpaccio of pigeon. Her feathers were a fractured mix of buff and cream with brown tones spread in tiny spattered ovals across a base of pure white. She was immaculate: no dents, bumps or broken feathers. The muscles of her chest are swollen with healthy fat, her feet tense and tight. Looking into her eyes, she lacked all definable human emotion, I saw no recognizable expression. She protected herself with a self-contained, seemingly immutable stillness. She was the perfect echo of a snow-covered prairie. I fell instantly in love.

In the truck she was folded gently into a tube of elasticated cloth, preventing indignant struggling and the breaking of feathers. On the weighing scales, she came in at just under two pounds. A healthy wild falcon, trapped using methods invented long before North America was 'discovered' – methods the tribal falconers in the Pakistani compound would recognize as their own.

Once trapped, a wild hawk or falcon needs to be trained. Like the little musket belonging to the boy and his father in Pakistan, or the falcon I trapped on the prairie, a wild bird of prey will naturally exhibit fear in the presence of humans. To create a working relationship, this fear needs to be eradicated. Almost a lost art in the West, 'waking' is one of the oldest known ways of overcoming the bird's fear. The process still exists in the most traditional falconry enclaves, and it is a method still used by the tribal falconers.

In the case of the little boy's trapped and sealed musket, the process of waking begins when the hawk is taken to a darkened room. The threads sealing the hawk are gently untied and the lack of light in the room acts like a giant black box, keeping it calm. For three, four days the hawk is kept awake, rotated, swapped and passed between participating members of the tribe. At around the third day the musket will enter a trance-like state, one that temporarily bypasses fear.

Once shrouded by this frame of mind, it is taken outside and exposed to various situations, without hysterical reaction. From this point the hawk is quickly trained to fly to the falconer, then to fly free and, finally, to go hunting.

As with trapping, the theory of waking is simple, the practice markedly different. The reality of waking is a quasi-mystical experience that takes almost a hundred hours, a process that stretches time and tiredness, producing a weird and improbable synaptic melding between the hawk and the human. As with sealing and trapping, waking has evolved out of a detailed understanding of a wild hawk's psychology.

I know this because I have tried it.

In contrast to the tribal falconers in Sindh, the goshawk used at the waking I attempted was bred legally in captivity. From hatching to rearing, she had no contact with humans. Removed from the chamber at roughly the same age as a trapping occurs in Pakistan, she was wild and stroppy. Weighing over three pounds, motivated by fear and anger, she was not easily handled and potentially dangerous. A fearful, aggressive female goshawk can pin your hands together; if she becomes attached to your neck, or face, you would require hospital treatment. Significant care and kid gloves are required during a waking, and previous experience with a hawk is essential. It is not a solitary pursuit; the inherent dangers and length of time involved preclude individual success. If a lone, inexperienced falconer tries waking, a slight madness and failure will follow. As a process, waking is a serious collective human endeavour.

We – four men and two women – started the waking in my darkened cottage. There in case of emergency or accident, among our group was an avian vet and falconer. We worked on rotation, hoping to keep the process as close to the methods of Pakistan as possible.

When scared, a hawk will puff up and spread its wings, doubling in size, becoming intimidating. Throughout the first twenty-four hours our hawk was agitated, frustrated and wild. She threw herself forward off the fist, hung upside down, her beak open, in a constant state of hostility. Each time she was gently lifted back onto the glove. All eye-to-eye contact was avoided and our movements slowed in lock-step with her mood.

By the second day her feathers were no longer puffed up, and she was no longer leaping regularly from the fist. All hawks have a thin, opaque third eyelid, a nictitating membrane. This membrane flips up and covers the eyeball when a hawk blinks. As the goshawk grew tired, the nictitating membrane lifted slowly; her outer eyelids, drooping to slits, were three quarters closed. She was floating softly between the fragile line of sleep and consciousness. When the outer lids touched together we all moved position, gently rotated the glove or swapped her to another falconer. Suspicious of change, her eyes would flip open, the nictitating membrane slide back below her lower lid, and she would remain awake and alert. By mid-morning of day two she had calmed considerably. We could move about without triggering fear and she slowly began to tolerate our company. By late evening, she gently entered a hypnotic, hazy state, as if trapped in a soft bubble, the volume of her world turned down and tuned out. The transformation was spellbinding; all residual fear seemed to leak away, disintegrate and evaporate from her feathers, replaced by an air of quizzical bemusement. She was awake but seemingly dreaming, capable of occasional alarm followed by longer periods of calm.

We were equally dazed. Some of us had slept, some just dozed, all hollowed out by time; conversation and excitement long since drained. We shared lasting moments of silence, gaping holes, repeated conversations, cyclical, a dull thudding in the head. Flashes of anger and annoyance were forced back, a missed fragment of meaning, a vague insult, perception unsure, remain calm. Those of us who had not slept entered a space both real and surreal, an unstable fine line between knowing what was factual and what was imagined.

The morning of day three entered darkly. A log fire burning, all lights off, in tired silence the collected group melted, blurred, obscure in purple shadows. As the light outside increased the atmosphere within changed. The hawk and the group parted, everyone now fidgety, but calm. We took a stroll across the fields through light mist rising. Small birds broke open the dawn, a mounting, stunning cacophony, loud across the cold and chill of the early hour. The silence between humans was deafening. At a farm gate a huge shire horse stepped forward with sweet breath snorting. Frightened of her own reflection a day before, the hawk leant forward and gently touched the huge beast with her beak.

In the garden of the cottage the hawk was finally released from the glove and placed on a perch. She took a bath and drank deeply from the water. To bathe in the wild puts a hawk in a position of vulnerability. They bathe only when relaxed or confident of safety. It was a good sign. When finished, she hopped back on to her perch. The sun was warm and cast fingers of light across the garden and over her back. Relaxed, she slept standing up for the briefest moment.

Taking her back up on the glove, something had clicked. We had pushed her through a barrier, and now coming back into focus, the hawk was entirely without fear. It was only then it occurred to me the term 'waking' stemmed not from the obvious, not from keeping her awake; rather, it was about waking her to a new state of being.

The sun was still bright as we walked back along the lanes then drove off in cars, ending up at a busy country pub. The hawk, perched happily on the edge of our table, became an obvious centre of attention. Children and adults came up, asked questions. Different dogs scurried about, eating crisps off the floor. At one point the hawk flapped and scattered drinks across the table; a crashing noise, splintered glass bottles spun on flint. Unfazed, she regained her perch on the corner of the table and watched as we cleaned up the mess. Later that day, she was happily driven home, and looked out of the car window from the glove. The waking over, the next stage of her training could begin.

The discovery of this gentle transition from fear to calm through shared sleep deprivation is one of the most interesting feats of our knowledge and understanding of hawks. Waking is unique among the ways humans have learnt to train any animal. It has also survived the test of time; it is no surprise then that the young boy with the trapped and sealed musket was being shown how to use this method by his father in Pakistan.

Back in the compound, it was now approaching mid-morning and the swapping of stories and discussions about training and trapping are over. Different members of the tribe wander off to farm the fields or work in a nearby seed factory. Chanesar walks across the compound and picks up his goshawk, motioning me to join him, and we walk to a wash of water behind the far wall. Three egrets, immaculate pure-white sentinels with lime-green beaks, point at the surface, fishing. The goshawk twitches. Stirred to flight, the egrets arch up into the sky. Chanesar slips his hawk, the clip of her wings triple that of the target. The slowest egret shrieks, twists and turns back to earth, and dumps on the ground. The goshawk skims above it, flares her tail, and shoots up vertically to about twenty feet, ready to strike. The egret, sensing a chance, seeks sanctuary beneath Chanesar's legs. The goshawk, confused, misses. The egret, panicking, flaps forward back to the water and attempts to dive under. It is too shallow and the goshawk pins and pushes the egret easily in to the mud. Chanesar removes the egret from the hawk and ties it to the smallest child among a group gathered watching us. The little boy, beaming, spends the next few hours carrying the egret around as a new pet. He touches and strokes the feathers and makes the egret fly before pulling it back to earth. He talks to it and wanders about proudly showing it off to anyone who comes near.

Just before we leave for the proper hunting, the egret is taken from the small boy by an older sibling. In front of his brother, the young man kills and butchers it, waving a leg and wing to taunt him. The little egret owner, bereft, hysterical, cries for a good fifteen minutes.

At the hawking grounds we quickly flush several francolin, but the goshawks are unlucky and fail. The last francolin flies across a huge distance and, following it, we cross a tarmac road. As we do so, Haider points to the sky. Above our heads, a wild peregrine, a shaheen, is halfway down a long, twisting, vertical stoop. It is accelerating at immense speed. Below it, a pigeon rolls, jerks, then drops. Together they disappear behind the far tree line, presumably a successful strike, as we do not see the shaheen or pigeon again.

Punhal walks over with his sparrowhawk to the spot where the francolin landed. Usually a hawk is flown from the fist, allowed to pursue the quarry under its own impetus. Sparrowhawks will naturally try to kill small birds, lizards and mice. It is very common for a sparrowhawk to lunge repeatedly at these creatures as a partridge flushes, this lost split-second handing advantage to the francolin, the flight a failure.

Being small and slight, sparrowhawks have high metabolic rates, are wound up tightly and ready to explode. In short bursts of intense speed, they cover one hundred yards in a matter of seconds. Continued bating and lunging wastes valued energy. In preparation, Punhal places his hand on the sparrowhawk's shoulders. The little hawk rolls over, almost like a dog having its stomach scratched. Punhal gently folds her into a patch of cloth. Instead of flying from the fist, he is going to throw the sparrowhawk like a dart. Casting and throwing is a longstanding technique, developed when landscapes were unspoilt and appears in descriptions contained in some of the oldest falconry books in the English language. Using this technique, Punhal has total control. He is able to select the correct quarry, prevent bating and conserve the much-needed energy for flights at francolin. Punhal is the only falconer I have ever seen attempt it. For such a highly agitated and complex bird, his sparrowhawk gives no indication of being upset or alarmed.

As I follow Punhal through the scrub looking for francolin, his small hawk twists her head around and stares straight at me. She is a calm, bug-eyed baby bird wrapped in swaddling cloth, it is an utterly surreal spectacle. Skittish birds zip back and forth from under the cover, Punhal resists, and we avoid many wasted flights at the wrong quarry.

We eventually scare the francolin, but it takes flight too far ahead, the distance well beyond Punhal's hawk. Once again we watch where it lands. Reaching the area, the falconers track its footprints in the dust, finally surrounding a small bush. With the francolin locked down, there is no need for throwing. Punhal gently unfolds the sparrowhawk. She rolls back over and clambers up his fist, standing ready on the glove. We swish at the bush, a flash of black, a cry goes up. The sparrowhawk sprints violently, striking the francolin hard across the back, and they land in a ditch. Punhal runs over, pushes his knife between the toes of the hawk, cutting open the francolin's chest. He invites me in closer. As I watch the sparrowhawk feeding she pauses and looks up from her kill. The sun now low, floods the land in blood-red and liver tones. Our day is over. My time is over. We reach the compound and load the jeep. I shake hands with Chanesar, Punhal, Ghulam and Haider. I see the shadows of several women, hidden and moving behind a sheet at the far end of the compound. The children run about around my legs. We take farewell photographs of each other. I am presented with assortments of cloth, a patterned blanket and a rare ceremonial leash. I thank everyone and say goodbye.

Salman drives out into the dark and approaching larger human populations, numerous roadblocks begin to appear. Armed men take bribes before letting us pass. From the impenetrable blackness, an unmarked white car passes, pulls in front, brakes hard to slow us down. The men in the back stare through the window, then the car accelerates away into the night. Salman motions to a compartment in the dashboard; inside it is a loaded gun. He smiles, winks and says, 'Any problems, Mr Crane, start shooting.' I am not sure if he is joking. As we continue on, for warmth, but also as a disguise, I wrap myself in the patterned blanket, leaving only a slit for my eyes.

Twenty minutes later, on the outskirts of a small town, the same white car is stationary at the roadside. Several armed men lounge around the bonnet, smoking and talking. After waving us to pull over, they empty the jeep, unpacking our bags and belongings. The telemetry is found and removed, they spread unfamiliar technology across the dust. A few terse words are exchanged and a hefty bribe placates the police. Within half an hour we are off the road and have found somewhere safe to stay. We will leave very early the next morning.

Long after dawn on the edge of Karachi we stop at a McDonald's for breakfast. It is a sad, disappointing return to 'civilization', marred by predictability, greed and intimidation. Not to mention slippery, synthetic-tasting chicken.

The Bell-Maker

The first call to prayer is droning out as we wander down a narrow side street in the city of Lahore. Sitting beyond an open door, smoking in his workshop, is a thin man of no more than forty. He is the bell-maker Mohsin Ali. Salman tells me his is a seventh generation of craftsmen; roughly calculating it in my head, this means his family have been making a variety of bells for nearly 300 years.

Mohsin spreads his tools out across the floor: a hammer, tin snips, a sheet of metal, round punches (large to small), a brass doming block, a flat file, flux, solder, tongs, a box of matches and a bucket of water.

Picking a flat sheet of brass, he cuts four circles of metal, each the size of a fifty-pence piece. They are heated with oxyacetylene, then plunged into water, and in the transition from extreme heat to sudden cold, the structure of the metal rearranges and it can now be hammered and stretched without splitting or tearing.

Taking one of the cut discs, Mohsin slips the flat metal over the top of the largest curved space in the doming block. With a punch and hammer, he taps and shapes the disc into the concave space of the block. When finished, the metal looks like an upturned contact lens. Selecting a smaller indentation on the block, Mohsin repeats the tapping and shaping. The smaller the indentation the more acute the curve of the metal becomes. The brass disc is less a contact lens now, more a large acorn cup. He repeats this process with the three remaining discs, then sets them in a row. Checking by eye, he looks for any discrepancy in height. The smallest and neatest is used as a template, the others filed to match.

Moving the domes on to a hand-sized rhino-horned anvil, Mohsin delicately taps the edge, creating a squashed band around the top lip of each cup. The four halves finished, he punches a hole in two of the domes then cuts two hexagonal pieces of metal of about the size of a dried chickpea from a rod of steel. He drops these dappers into the two punched half-domes. The un-punched other halves are coated in flux, then squeezed together with the punched halves, sealing the dapper inside.

Mohsin then cuts a strip of brass and bends it into a C shape. On the opposite side to the punched holes, two dabs of flux are smeared on the outer end of the bell. The small strip of C-shaped metal stuck to the flux becomes a cuff to attach the bell to the hawk's leg.

Finally, a long strip of metal is wrapped over the cuff and folded around the bell. The ends are twisted tight to hold the bell together. Mohsin places the bell on a heat-resistant tile and the oxyacetylene is lit and gently moved across it. The metal surface changes from gold to a deep orange; flux bubbles and steam blows out from the punched aperture at the base of the bell. Without this, the bell would explode under the pressure of steam burning away inside it. The bell now glowing red, Mohsin dabs silver jeweller's solder across the join and it disappears like quicksilver. I smile. An 8,000-mile round trip for this single piece of information.

I have been using the wrong solder.

Mohsin lets the bell cool naturally. The soft metal strip holding the bell together is removed and a hollow, dull, clonking noise emanates from inside. Mohsin picks up the bell and, pulling a hacksaw blade across the punched hole, creates a thin slit across the bottom of the bell. Instantaneously, it comes alive with a mid-level pitch of great beauty.

The whole process has taken an hour. With both bells complete, Mohsin gives them to me for free. They are scuffed and pitted, coated in a dull orange patina. They look like they have been forged and rolled straight out of the earth, a beautiful set of pock-marked pebbles taken from a beach. Handcrafted, totally unique, tone perfect, these bells have both individuality yet encapsulate a story of historical significance.

A set of uniformly machined falconry bells will cost twenty to thirty pounds in the West. The handmade bells of Mohsin Ali sell for less than three. A day later I flew home, taking hundreds of bells back to England, selling them for eight and ten pounds a pair to as many falconers as would have them. All the profits were wired directly to Salman in Karachi.

A few months later a stranger arrived at my door in a black Audi. He asked questions and took photographs. The frequency and level of payments had aroused the interest of MI6; the tacit suspicion was that I was funding a terrorist organization. No doubt my name and address are still on file – a fact I am hugely proud of.

2

Further Travel

Although the images remain vivid, it has been eight or nine years since my trip to Pakistan. It is late summer at the cottage and the evening sun creates a muggy, warm, slightly sticky but comfortable atmosphere. The wild rehabilitation sparrowhawks in my possession are currently settled in their respective aviaries for the evening. I have done all I can to help them. In a few days they will be released.

Half a mile away from the cottage is a neglected tract of old common ground abundant with wild grasses and flowers. With nothing left to do today, I take the dogs for a walk and, reaching the field, I swish out to the centre and lie down, partially hidden in the itchy cave of grass. For ten monumentally annoying minutes the dogs jump on me, roll about, fight, fidget, then sigh, succumb to the heat and finally settle. Gently dozing, I close my eyes and listen. Initially, the noise is the simple somnolent hiss of wind through grass. Slowly, the sounds of different and distant birds tune in. The searing squeal of swift and swallow, the ocarina blowing of wood pigeons, the squeaking-wheel *pwee* of an oystercatcher on its way to the reservoir. I hear skylarks, pipits, a cock blackbird and the *eek-eek* cry of kingfisher scooting along the brook. I count a dozen different birdsongs, all in whirring clicks, beeps, squelching chatter and high held, sharp screams. After an hour or so the noise fades then ceases abruptly. I open my eyes, stare into the sky and wait. The peregrine arrives. When I see her, she's coming in low, low enough for me to see her in intimate detail. Her wings are pale, long, pointed and move in shallow, pulsing beats. Bars of muted grey streak across a cream chest. A bandit mask of slate blue drips down around infinitely dark eyes. She spots the dogs, twists up and skirts wide, a spiralling circle mounting fast, rising higher, smaller, a dot, then gone.

Although brief, it is always enough. The fastest living creature on the planet has just passed across my tiny sphere of consciousness, blessed me with her presence.

This particular field has witnessed more than peregrine falcons. It has significance and holds many memories. From delicate merlins to heavy-set, hard-hitting goshawks and slight, bullet-speeding sparrowhawks, all have been trained by me in this field. It was in this field, three years before Pakistan, in the summer of 2004, that I trained and flew my very first hawk, a Harris hawk called Cody.

Imported from America to the United Kingdom in the 1960s, Harris hawks became popular very quickly. Unlike goshawks and sparrowhawks, they are gregarious, easy to train, easy to breed, not particularly demanding, require low-level skills and a moderate sense of detail in the falconer. This ease of training makes them the perfect hawk for a beginner.

In the woodland near this field Cody and I had a lot of adventures, successfully catching rabbits and, occasionally, pheasants. Surrounded by shoulder-high bramble, deep within thistle and dead nettle, he would pivot up and perch on the branches above my head, scanning the ground for movement. As I pushed and bashed through the intimacy of heavy cover, he hopped from tree to tree, waiting for quarry to flush. On a dark December morning, rising off the damp ground with a jerking, flicking wingbeat, he caught a woodcock on the rise. In my hand she was soft, a beautiful bronze and copper feathering against leaf litter, with a long, strange, soft, grey beak: a keratin-coated straw perfect for sifting molluscs. I let him eat half, then flash-fried the remainder when I got home. It was the first woodcock I had ever tasted. It was an incredible meal.

In the very same wood I once narrowly missed stepping

on a newly born fawn. Still mucus-wet, perfectly still, it was curled cat-like in the centre of a flattened patch of grass. No bigger than a young hare, the fawn would have made an easy kill. Instead, I called Cody down, secured him to the glove and walked away. As we turned on the outer edge of the nearby fields, his mother briefly marked the skyline before slipping back inside to join her baby.

In heavy winds, on the steep slopes of snow-covered hills, I let him rise off the glove, twist up into the air and soar high above my head. Trained to remain in position, he would stoop like a falcon when rabbits were provoked to run free from heather or bolt from buries. He was an extraordinary hawk in more ways than one. So enamoured by his ability and behaviour, he inspired me to travel and opened up a world more nuanced and far reaching than the simple satisfaction of localized hunting experiences.

Harris hawks

In the wild the geographic range of the Harris hawk is large, extending from the tip of South America up through Mexico, Texas and on to California. Across thousands of miles of varied terrain, Harris hawks are ubiquitous, successful, apex predators. To look at, they are fairly plain: generally about the size of a European buzzard, coloured a dull rust brown in maturity, with a broad flat head and cream markings on the tips of their tails. Their legs are strong and long, a catwalk model's length of leg, longer than in many other hawks. Their wings and tails are broad and tough, withstanding extreme variations of heat and cold. They have a natural aptitude to spiral up on thermals, scoping out suitable prey over vast distances.

In the enclosed spaces of cacti and desert scrub Harris hawks are capable of short, slow, stabbing flights at their chosen quarry. Intelligent and opportunistic, they will kill small mammals, take chicks from nests, scoop up snakes and scavenge carcasses. What they lack in speed, guile and aesthetics, they make up for with cunning. Harris hawks are one of only a handful of hawks that either live or hunt in extended groups. For a hawk to hunt with extended family members co-operatively, in pairs or as a shared collective pack, presupposes some form of communication, a reflexive intelligence – even a form of cognition. It was my inquisitiveness about Cody's evolved history and his biology and my desire to witness Harris hawks in their indigenous landscape that took me to Texas in 2005. This in turn triggered my trip to Pakistan, and then an obsessively intense period of travel that was to last four or five years.

～

Texas

On the furthest southern border of Texas, a stone's throw from Mexico, temperatures in the summer rise to 40°C by late afternoon. It feels drowsy, still; the heat is gruelling, the atmosphere silent. It is so hot most Harris hawks are under heavy shade and, so far, I have failed to find any. Those braving the heat look like mere specks of grit, hundreds of feet up in the sky.

On the roads, bubbles of tar ooze black and tacky soft. Cars have killed wildlife; the smashed flesh of snakes, javelina (wild pigs) and armadillo feed vultures, caracara and the maggots of a thousand flies. Turning off the main roads and traversing the edge of the Gulf of Mexico, driving on disused dirt tracks, the land is flat, arid and complex. Huge mesquite trees and cacti are dense, much bigger than expected. Individual cacti pads are as wide as a tennis racket. Fifty or sixty cacti rackets collect together in a single lime-green plant the size and shape of a deflated hot-air balloon. Across their turgid surface ants and insects dodge between delicate flowers coloured sun-yellow and cadmium-red with ivory spines several inches long. Trapped under the cacti, or tumbling over the dirt, knotted balls of weed blow about in a breeze rolling in from the ocean. A roadrunner, a pheasant-sized brown bird, streamlined, slips and slides between thin blades of dry grass then sprints parallel to the car for thirty yards. A flock of wild turkeys launches from the side of the road, rolls across the bonnet; some bounce off the window, nearly cracking the glass.

By late afternoon, as the sun loses intensity, I find my first family unit of wild Harris hawks. Three juveniles are perched together on a telegraph pole. I stop the car almost underneath them. Their attention elsewhere, they remain on the pole, looking out over the cacti. Floating in low, slow circles, I watch two large females swirl over the ground. The lowering sun casts liquid shadows across the earth and the shifting silhouettes of airborne hawks spook a small rabbit. It panics and scoots across an open area of sand. One of the juvenile Harris hawks drops off his perch, jags down between the cactus and pins the rabbit under the front porch of a derelict house. In quick succession the rest of the unit follow in on the kill-taker, scrabble in the dust, flaring wings and feeding. I watch contentedly, then start the truck and head back to the motel.

Just before dusk I take a walk, turn right out of the motel towards the Mexican border. In the distance, neon signs and billboards wobble in the spectral early-evening heat. There are no pedestrian paths; everyone is forced to drive. Unable to cope with the mounting fumes, and afraid of being stopped by the police or hit by a car, I step off the road. On the ground, presumably cast from a car and wrapped in cellophane, I find a bag of rock cocaine the size of a baby's fist. I pick it up, toy with it, then drop it into the grass. Inside a chain-link fence is a dead and disused oil well. The ground is littered with discarded tools, food wrappers, empty drums, rusted metal and iron sheeting. I start lifting rocks for fun and find translucent yellow scorpions. A fat black spider rears up with lifted fangs, protecting eggs in a soft white ball of web. Clattering a sheet of metal on rocks, I startle a covey of scaled quail, and the noise sends them on flickering wings, bickering up over the low brush. In a sinkhole a spread of delicate flowers scatters like confetti, rolling across the sandy soil like puffs of purple orchids. Even in this industrial landscape, these pockets of wildness are waiting at the edges, patiently waiting to return properly.

The next morning, long before the heat takes hold, I pack up and drive to a new motel 200 miles away. On the back roads near the towns of Freer and Hebbronville, I find a lone male Harris hawk on a low branch. His head bobbing, his gaze transfixed on something hidden in an adjacent tree. I stop and watch. I recognize his behaviour: just like Cody, he is working an angle of attack, waiting for something to flush. Through the mesquite he disappears from sight. An explosion of yellow orioles react, shrieking, lassoing and looping above the hawk in the light blue sky. Unperturbed, the hawk curls around, rises up then lands on to the top of a tree. He begins to pull their nest apart, attempting to get to their chicks. The orioles, pushed beyond the limits of fear, strike the hawk on the head, pulling feathers free and forcing him from the nest site.

Later in the week, I stop for a day at a vast inland lake and fish for bass. The evening chorus of grackles and small wading birds' whines and hisses sails through the air like phosphorescent, crackling static. Having caught a big fish, I cook it on one of the outdoor barbecues dotted about the park. The clean flesh, fresh, white and steaming, tastes like perch or cod. I carry a Lone Star beer and walk the empty shore. Finding a baby alligator sitting on the water's edge, half in reeds and half out, I foolishly reach down to touch him. His mother (the size of a sofa and perfectly hidden) explodes out of the lake. I take to my heels, and the shocking squirt of adrenaline makes me shake and laugh uncontrollably for half an hour.

Driving to the airport on the last day, I pass a desert correctional facility, a prison. I stop and hop from the truck. Walking along the roadside ditch, I end up parallel with a Harris hawk's nest. It looks an easy climb for a nosy human. For twenty minutes I try every conceivable angle, at each turn pushed back and speared painfully by chest-high cacti. A perfect piece of natural history: the nest flawlessly protected by a dense circular wall. It is a clever, well-considered design, the placement of nest and jail perfect for keeping coyotes out and dangerous men in.

On my return from Texas to England something shifted. Having seen the way in which wild Harris hawks fit perfectly into a specific ecosystem, seeing the delicate structure they had evolved from, noting their specific patterns and typical natural behaviours, something vital now seemed missing every time I took Cody hunting. At the time I was unable to pinpoint what it was. It was not until eating spiced francolin in Pakistan, and when I showed Haider some video clips of Cody hunting, that the penny dropped. Watching Cody swirl about on the little camcorder screen, Haider was polite, impressed even, said Cody looked like an eagle. He wondered aloud, asking why it lived on the land in the film; it did not seem to fit. He asked if the wild hawks we have in England are not good enough for falconry. I explained that, under English law, falconers are no longer able to harvest or trap wild hawks. A knock-on effect being the development of captive breeding and the importation of birds of prey.

I explained that Britain has hundreds of licensed breeders, all supplying a range of non-indigenous hawks, eagles and falcons imported from across the world: New Zealand falcons (that's their name) from New Zealand, Aplomado falcons from South America, bald eagles, red-tailed hawks, Harris hawks, diminutive kestrels and prairie falcons from the US; lagger, lugger, lanner and saker falcons from the deserts and continents of the Arabian peninsulas; pale-chanting goshawks from Africa, black sparrowhawks from South Africa, bi-coloured sparrowhawks from Peru and shikra from India and Pakistan. Non-falconry exotica also abound: palm-nut vultures, tropical screech owls, Turkmenian eagle owls, burrowing owls, Indian scops owls, Magellan eagle owls, caracaras, Malay brown wood owls, jackal buzzards, kookaburra and Harlan's hawks. On top of this we are also able to produce a wide range of synthetically-bred hybrids, whipped up and criss-crossed in a glue of semen and egg. The peregrine falcon mixed into a merlin, hybridization in name and body: a perlin. Gyr falcons crossed with merlins and prairie falcons. Peregrine, sakers and kestrels crossed with merlins. All shifted about through artificial insemination, artificial falcons with artificial names, the gyrlin, the peresaker, the gyrprairie and the kerlin. What started with falcons has spread to eagles and hawks. European goshawks crossed with American Harris hawks and sparrowhawks, golden eagles crossed with red-tail hawks, indigenous with imports, mixed and matched solely to scratch the itch of human curiosity. Haider's innocuous question 'Are the hawks you have in England not good enough for falconry?' was as simple a deconstruction of the situation as it was possible to have, and it remained with me all the way back to England. What had been sewn as a seed in Texas, what transformed my thinking in Pakistan and eventually led to the rehabilitation of injured sparrowhawks in England, is that the true power of birds of prey (and, by extension, all falconry) stems from a direct connection to specific landscapes and the quarry they support. No matter how much I tried to convince myself to the contrary, an imported hawk such as a Harris hawk would only ever take me so far. Flying and owning Cody began to feel like a false premise. Moreover, if he escaped or was lost, I would have inadvertently introduced an invasive species with the power to survive and the temerity to kill and displace indigenous hawks. This was a shocking realization.

The choice was complex. His wellbeing was paramount. Even if it was legal to do so, for obvious reasons I could not release him. I did not want to sell him (he was more important than money) and I certainly did not want to keep him in captivity like a pet. What he needed was a place where he would be treated with the respect he deserved and flown in exactly the same manner as I had been doing for the last two years. Inevitably, the solution came from outside England.

Croatia

Now firmly established on nearly every continent, falconry is practised in at least sixty-eight different countries. High-speed digital communication on top of the human need to organize and connect has created clubs and organizations in almost all these countries. Falconers the world over communicate via social media, and through the most significant club of all, the

International Association of Falconry (IAF). Having used these two sources both to research Texas and to make contact with Salman, I had cast my net wider, seeking an education well beyond my experience, to connect with people who lived their entire lives vicariously through birds of prey, to strangers with the good grace to mentor me and play a part in firing my desire to fly purely wild indigenous hawks. And it was one of the falconers in this wider net who offered to provide Cody with a new home. So, a few months after my return from Pakistan, I set off across Europe to Croatia.

It's late summer of 2007 and I am sitting in a café in the Croatian town of Karlovac. The architecture across the street is odd. Deep, pitted scuffs, gouged, chisel-like wounds, large circles and smaller chipped dots seem to have been spat across the walls of the buildings. High up near the roof, one hole cuts straight through thick alabaster plaster to orange brick. The marks are beautiful. Like air bubbles frozen in stasis, or raindrops on dry dust, it looks like the surface of a full moon on a clear night. Their strange beauty belies darker origins. Each pockmarked scar was made by bullets, mortars and shrapnel as soldiers fought their way towards Zagreb in the last Balkan war. When Croatia was piecing itself back together, the man I am due to meet was a young boy and a passionate falconer. Scrabbling about in the wreckage of bombed-out buildings, looking for gloves and falconry equipment, his neighbours laughed and told him he was mad. But Viktor's madness never left him. The fizzing hyperactivity and passion of childhood has been funnelled into writing a PhD paper on wild grey partridge, and breeding both highly sought-after goshawks and some of the finest pointing dogs in the world. He is also at this time the IAF representative for Croatia.

Britain has thousands of Harris hawks; in 2007 there were less than twenty in Croatia. At Viktor's request, I have driven Cody across Europe and delivered him to Viktor's friend, a man named Christian Habich, in Austria. Cody will remain with Christian until his export paperwork has been cleared, then he can be imported legally to Croatia. Before making the final leg of my journey to Karlovac, the last remaining image I have of Cody is a calm, handsome Harris hawk sitting in the dappled morning sun of a lush Austrian garden. He rouses and begins scanning the hedges and apple trees for movement, looking neither concerned nor remotely bothered by my departure. His perfectly evolved lizard-like mind is well beyond sentimentality and the type of attachment I feel. I take a few pictures, wishing him luck, and thank Christian. I leave knowing I have done the best job possible; that once in Croatia, under Viktor's guidance, he will have a life arguably better than the one he has had, or would have, in England.

~

During the torrid heat of a Croatian summer, one of the least-known migrations of birds takes place. Quail are small, no more than the size of a human hand, and speckled light brown with flashes of pale ochre, black and grey. This dull camouflage helps aid survival against predators, but it is the speed of their flight that makes them particularly special for falconers. Quail are a migratory indigenous gamebird on a par with the francolin of Pakistan. For thousands of years they have flown through Turkey and across the Balkans, taking up residence in eastern Europe for the spring and summer. The wild sparrowhawks and goshawks in these areas make full use of this fresh supply of food. A handful of Croatian falconers have learnt to do likewise.

It is unusual to hunt with a goshawk in the summer, particularly in temperatures of 30 to 40°C. The heat causes the metabolism of the hawk to slow, and they do not respond well. It takes very capable and experienced falconers to get summer hawking right. More often than not, they fail.

Viktor wakes me at 4 a.m. after my first night in Karlovac. I slide out from under a thin cotton sheet in the heat of the night; the sweat from my skin has turned it opaque, the sheet now sticky, uncomfortable, like butter-covered baking parchment. Viktor is bouncing about with energy and focused fun, his generous enthusiasm undimmed even by the early hour.

His hyperactivity is infectious, and I pull myself together and get up. Tim (the goshawk) and Ella (the pointer) are ready and waiting in the car. Viktor's mother brings a small, intense coffee, which we follow with a chug of iced water, and we are on our way to the hawking grounds.

The countryside around Karlovac is one of the least spoilt landscapes in Europe, a mixture of low-level farming and wild spaces jumbled together in a mishmash across lush, fecund and beautiful land. We get out of the car and walk through thick summer growth, but tracking quail is difficult, distraction easy. The fields are buzzing with life. Insects are hopping, jumping, creaking and noisy. Darters and dragonflies clatter and spin about, catching light on wings like droplets of white fire. Working busy-bees fumble into purple jug-shaped flowers then back-out with bums covered in pollen. Big spiders, half the size of a hand, shake on webs between tall grass. Their soft, fat abdomens are covered in flashes of lightning yellow, and shiny black patent-leather legs entice me in close, daring to be poked. On the floor armoured bugs are bumping, little green, brown, black and rainbow tanks stomping with micro-footprints over patches of dry soil. A guinea pig on stumpy legs shoots through with a puff of dust into a grass tunnel. Waist-high wild grasses and seeds stick to cuffs, early-morning dew soaks through to the skin and my feet are sludgy, wet and warm. Spears of bright yellow flowers tower high over a swatch of cornflower blue. There are delicate red, pink and orange patterns, subtle green stalks arrive in a multitude of gradations and tones. The purple bloom of distant mountains rises up to a cracked thin black line spreading the width of the horizon. In the valleys, rivers and streams flow clear and warm. We find a deer lying in the shade of a disused hospital, then spot a wild goshawk. A sparrowhawk and a peregrine falcon follow in quick succession. Viktor tells me that further north, deep in forest enclaves, lynx, bear and wolves can still be found.

We get two chances at quail. Ella weaves left then right, working out the scent. The quail does not wait for the dog to arrive. It shrieks into the air with preternatural speed, rising high and far ahead, and the distance between quail and the goshawk opens up like a cut across a drum skin. The flight is finished in the blink of an eye. The second quail is just as fast. The dog approaches, pauses, moves forward, then locks solid. We move to the left. The quail bursts from the ground, straight towards me, curves round my shoulders on thrumming wings through the air, beating Tim completely. Quail are simply astonishing.

By 8 a.m. the temperature is rising up through 30°C. It's too hot, so we stop and drive to the rivers of the Duga Reza valley. Near an old mill, flowing under a rickety bridge, water runs with a tropical green-blue tint. Diving through the water with goggles I swim over shallow runs of gravel, clusters of minnows; a flotilla of silver torpedoes peck and mouth drifts of food. Deeper, in forty feet of water, curved bowls of stone have been eroded over time and are surrounded by thick blankets of underwater cabbage. A place where pike and perch lurk, ready to strike. Numerous species of fish – chub, roach, barbel and trout – swim about, feeding freely. One or two vibrate alongside one another, laying eggs under cloudy puffs of milt. A water snake slips into the stream, moving through the surface water with a quick, curving motion. I pull up parallel with his bootlace-thin body. He turns, bumps against the tip of my nose, big, disc-shaped eyes set central in a slender head the size of my thumb.

In the late afternoon, bored with swimming, I walk to a nearby bar and order bear steak with cranberries. Two hours later, full of wine and false bravery, I set off on an extended expedition, floating miles downstream on an inflatable bed. I cast a fly rod and line at shadows near sunken logs. At the furthest point I dare myself to try a steep waterfall, but the boat tears and deflates on sharp rocks and I sink mid-stream, losing my rod and reel to the slow, deep depths. The swim to shore is easy. I pull the dead plastic up on to the bank and flop down, stranded in paradise.

The next day, I meet the future: my first rehabilitated wild sparrowhawk. She has fallen from a nest felled by loggers in the mountains; the rest of her siblings were killed. Partially feathered, she has been taken into the care of Viktor's friend, a former soldier called Zlatko. Viktor and Zlatko name her

Bok ('hello' in Croatian). Bok is kept indoors around humans twenty-four hours a day and is as tame as Punhal's sparrowhawk.

We sit around in the shade talking and handling 'misplaced' and 'lost' weapons of war. Rolling 9mm bullets back and forwards across the table, Zlatko places Bok on a perch. I cup her between both hands and feel her heart pumping in frenzy. I blow on her head. She remains resolute, unfazed, bored even. This is the first time I properly notice the smell of a healthy sparrowhawk, and the first time I have ever handled one.

Before meeting Bok, my first introduction to sparrowhawks was delivered fast and first-hand by nature, but at a distance. Packing and preparing for Pakistan, a gun-sudden shrieking splits the moment in two. Looking through the window, I watch the zinging black tracer of a frightened blackbird drop over, down and along the garden path. The sparrowhawk flicked up and arches over the burlap fence, intent on killing it. A split second's silence, and, in imagined slow motion, she misses. Time winds up, and both birds crash full stop into the ivy running up a nearby wall. A momentary pause, the blackbird taking flight first. Five feet out, and sensing death, he spins on a pinhead as the sparrowhawk cuts across his path. Both birds flip over and back to the ivy. The blackbird, instinctively clever, ladders down inside the bush and escapes. The sparrowhawk, distracted by my presence, wings splayed across green, looks straight at me. Then is gone. The speed, tenacity and skill of their flight was mind-blowing.

On my return from Croatia, as if prodding me into action, teasing and testing me, a second, more significant sighting of a sparrowhawk occurred.

For nearly 400 years, farmers, carts, clergy, country folk, horses and hawks have used a small bridleway near my village. Midway along its length the undergrowth curves tube-like, a twisting mess of bramble and high-sided trees. Firing through the wood, up across a stream, sparrowhawks corral and startle small birds into the tube. With no angle of escape, they are plucked deftly from the air. The only traces left are feathers spinning lightly through shive sunlight.

Once when I was walking in this tunnel, I saw a female sparrowhawk pinning a pigeon to the floor, silently tearing at flesh and feathers. She was so focused on her meal I almost stepped on her. Three or four seconds passed before the dog ruined it and ran past. The sparrowhawk flew off, leaving the pigeon. I lifted the carcass from the ground, brushed off the twigs and leaves and put it in my pocket. Back home in the kitchen, I cut the breast meat from the chest and peeled the legs from the carcass. The meat was still warm and soft. I dropped two bits to the dog; the rest went into a skillet. Thirty seconds raw to rare: nature's fast food delivered to me by an indigenous wild hawk. It was hard to ignore the signs this time.

For a falconer of my mind-set and thinking, the European sparrowhawk is the personification of the perfect hawk. Ubiquitous, supremely adapted, they are ferocious hunters, evolved to maintain the natural avian order. Short wings and a long tail provide high-speed manoeuvrability, and the twisting flights required in dense cover make them unparalleled hunters through tight woodland and along hedgerows. They have lightning-fast reactions and their eyes are an adaptive miracle, allowing them to pursue all targets with tenacity and guile. The underside of their feet is a mass of long, undulating ridges, the perfect glue for the feathers of any quarry. Whip-smart and opportunistic, to conserve energy they will drag larger prey into garden ponds, drowning them before eating them. Once, while I was driving a long, straight country lane, a female sparrowhawk skimmed out of a tree and flew six inches above the car's bonnet. She was cunningly using the sound and movement of the car as cover to flush birds from the hedges. It worked. She hammered left on to a flock of unsuspecting starlings feeding in winter stubble.

To look at, sparrowhawks are comically small, a cartoon character imagined and built by a child. A female's legs are no thicker than a pencil, the male's toes and talons the size of toothpicks. Seemingly too small to be dangerous, it is tempting to hold a male sparrowhawk without a glove. When it is clamped down in anger, a musket's back talon will glide smoothly under a thumbnail, causing electrifying pain and felling a ten-stone man.

In falconry terms, sparrowhawks hold a strange position. Historically only flown by women and the clergy, they were of no consequence in the masculine world, where goshawks, falcons and eagles were the measure of a man's power and status. Yet one of the earliest falconry books ever written in the English language concerns the training of sparrowhawks. Nearly 500 years old, *The Perfect Booke for Kepinge of Sparhawks* was found sealed inside the walls of a sixteenth-century house. Perhaps written by a woman or a lowly yet literate falconer and secretly kept, it remains without an author or any identified provenance. Marginalized and inconsequential in the patriarchal world of medieval royalty yet worthy of words in a time when few could write, sparrowhawks are a feathered conundrum.

Even in modern times, very few people have the nerve to fly them consistently. For hundreds of years they have remained an outsider's hawk, existing as a subculture within falconry. Owning them explains why. Training a sparrowhawk takes extreme levels of commitment, focus, time and risk. There is absolutely no margin for error. They die quickly, are lost quickly, are fragile, unforgiving and highly erratic. However, once you know them and have seen them working, sparrowhawks are beguiling, existing in an experience uniquely their own. They touch the highest reaches of technicality and style and fly with a hedonistic commitment well beyond their size. Weight for weight, sparrowhawks are one of the most significant indigenous birds of prey used for falconry anywhere in the world.

From the point I picked up then cooked and ate the pigeon in that lane, many sparrowhawks would enter my life, and I discovered a deep affinity for the species. The knowledge accrued from flying them filtered through to the eventual rehabilitation of their wild, injured cousins and, while I was training, learning about and flying sparrowhawks, I continued to travel, and to widen my experience of other species, other falconers and other landscapes.

Austria

On the way to Croatia and during the two-day stopover in Austria with Christian he discussed a project he believed to be one of the most significant historical developments in modern falconry. Those who work closest to nature are the first to see the destruction of habitats and notice environmental changes. Falconers, like the landscape and birds of prey, feel the pinch of increasing human populations directly. In many repressive and oppressive countries, and even in those considered liberal and democratic, new laws and political gain are used (often inadvertently) to restrict access to land, to quarry and to birds of prey themselves. We often have no option but to acquiesce.

A tall, precise and serious man, Christian worked with the other IAF delegates to draw up paperwork which would have falconry ratified by the United Nations Educational, Scientific and Cultural Organisation (UNESCO). Falconry would then be legally protected and classified as an intangible cultural heritage, and therefore exist outside laws which might ban it as a 'blood sport', or see it eroded by legal loopholes regarding quarry and landscape. The UNESCO bid was an impressive piece of forward-thinking legislation. Other parts of this project were to stage the largest festival of falconry ever held in the world, with representation from all the members of the International Association of Falconry, and to publish a complete photographic history of falconry, with contributions from each of the sixty-eight nations in the IAF.

While outlining the UNESCO project and plans for the festival, and the book, Christian drove me to the Austrian–German border and introduced me to two of his friends, who also happen to be two of the most influential eagle falconers in Europe, the husband-and-wife team Josef and Monika Hiebeler. Within twelve months of my return from Croatia, I flew back to spend time with the Hiebelers and their eagles.

Eagles

Eagle falconry is old.

In Kazakhstan, there exists an ancient bronze sculpture. Estimated to have been created in 2500 BC, this image is one of the earliest pieces of evidence of the existence of falconry in any form. It portrays an eagle falconer or *berkutchi*. The Kazakhstani *berkutchi* were some of the first humans to use wild-taken youngsters and trapped juveniles. They employed waking and other methods that not only informed the tribal falconers' methods but every single falconry tradition across the globe. In this sense, eagle falconry is considered by many to be the source of *all* falconry.

In contrast to its well-established and lengthy history in Kazakhstan, eagle falconry in Europe is a relatively modern phenomenon. In his book *Hunting Eagle: The Development of German and Austrian Eagle Falconry*, the falconry historian Martin Hollinshead gives dubious honour for the rise of eagle falconry in Europe to Adolf Hitler and the Nazi regime. On the cusp of the Second World War, the political spin doctors of the Third Reich (like other despots, political leaders, kings and queens throughout history) co-opted the eagle and other birds of prey as symbolic metaphors in their lust for totalitarian power:

> Under the Nazi regime, German falconry flourished; there were Gyr-falcon collecting missions to Greenland and Iceland, huge field meetings, and at the monumental Berlin hunting Exhibition of 1937, Germany's falconry was floodlit for the whole world… under the Nazis, Germany was also able to boast the building of the great Reichsfalkenhof

– a mews (a large aviary) – containing what were surely the most impressive hunting hawks anywhere on the planet... the eagle was a national symbol, its historical might and power flying from banners and looming from buildings across the land. Now at the Reichsfalkenhof, this fabled and fearless warrior eagle, the bird of myth and legend, would be bought to life.

When the war ended, the eagle, emptied of symbolic racist hate, remained as a hunting bird. The traditional gatherings continued too. It was at these gatherings, from the 1970s onwards, that Josef Hiebeler's methods for flying eagles became influential. These huge field meets stretch back well before the Nazi regime, and have always been formal events, with many falconers wearing the traditional dress, hats and jackets of their respective countries. Before meeting the Hiebelers, I attended one on the Slovakian–German border to see golden eagles hunt deer, to gauge the Hiebelers' influence and to pay respect to the *berkutchi* of Europe.

Slovakia

The temperature of the little wooden hut is a long way below freezing. For a few days the sky has remained aluminium grey as ice crystals float about in thick mist. Snow begins to fall in fat discs, forming fresh layers over drifts higher than a human. As I mooch about outside in the mornings, collecting wood or drawing water from the well, the clear scent of pine sap seeps from claustrophobic woods. Snow creaks underfoot and shafts of weak sunlight fall across branches then bend up off the snow in reflected pinks, mustard, blues and greens. Weaving drunkenly through the trees, the tracks and the footprints of hare, fox, deer, mink and pine marten indent and remain in the snow. It is the perfect place for eagles and there is a raw feeling of splendid isolation.

The cabin I am in rests on the edge of a small village. It is a close-knit community that calls to mind the darkest eastern European folk tales. Poor people not smiling give hard stares to strangers. There is no mains electricity, only a small log burner emitting bare heat and very little light. On the table a half-empty bottle of pure fruit alcohol is helping dull the brutal, relentless cold. The air in the cabin is foetid and stale. So far, there has been no movement or hunting and we wait for the weather front to change.

Mid-morning the next day, in a small community centre, an old, single-storey communist building, the atmosphere is thick with cigarette smoke. Cheap Formica tables are spread with a selection of heavily spiced, dry-cured sausage, bread, pickles, schnapps, slivovitz and beer. Thirty or forty falconers gather along the tables, eating and talking. There are a variety of goshawks and falcons on their gloved hands. In the far corner, on portable perches or nonchalantly placed on the backs of chairs, are four or five male and female golden eagles. This is the first time I have seen eagles up close and, in the context of the room, their size is overwhelming. I ask questions. Translation withheld, or non-existent, the *berkutchi* simply nod and point. Permission given, I wander over for a closer look.

Some creatures are undeniably big. When open, an eagle's wings stretch six feet or more. When the wings are folded tight and resting against their sides, these eagles stand four feet tall, their shoulders as wide as the trunk of a human torso. Their legs are almost the width of a human wrist, their feet and talons span the diameter of a small dinner plate. The largest eagle must weigh between twelve and fourteen pounds. I stand three feet away. Any closer seems rude, an invasion of her personal space. I can feel a tangible atmosphere around each eagle; it is tense, has the sulphur taste and humming noise of a million volts running through damp cables. Their magnitude seems to go beyond the observable measurements of size and weight – their true force lies outside the obvious. It exists in my mind as the silent potential of the unimaginable, unwavering damage they can do.

Over the years I have been attacked and injured by all manner of hawks and falcons. Not once did I feel true fear. I knew I could control the boundaries. This feeling, a benign, subconscious confidence, evaporates in the company of eagles. They are hooded and calm, and I know they are well trained and do not present a threat, but they still expose my vulnerability and lack of confidence. They force me to confront the softness, fragility and smallness of my own body. Without exception, each eagle is a withering presence dancing on the outer edges of extreme violence. A lion or tiger with wings, dressed by nature in a delicate bronze-crested necklace and a golden crown of filigree feathers.

As the winter festivities continue, the eagles become restless and slowly transition into *yarak*. One or two begin to call. A strange *chupp*ing sound bounces off the walls, high in tone, oddly delicate for such a large animal and feminine in pitch. The huge female moves position, slips, then re-grips the chair; a spiralling corkscrew of blue plastic peels to the side of her talon, leaving a deep groove.

From the community centre the various groups split up: falcons in one, eagles in a second, goshawks in a third. I remain with the eagles as we make our way to a bridleway adjacent to a pine wood. We are surrounded by wide agricultural land, a long way from the main village and other humans. By necessity, eagle falconry requires space. Their size dictates that all elements are amplified: the distances flown, the size of the animals hunted, the noise, the atmosphere and build-up all escalated to the largest possible scale. In preparation our group spreads out in a line a hundred yards long or more, and we begin walking forward as one. Downwind, several deer, startled at our scent, begin to move, jinking, criss-crossing fallen trees, leaping on spring heels out through the woods and across the snowy stubble.

An eagle is released. Her speed is deceptive, a slow, unfolding burn. She looks like a swan or heron taking flight. She moves as if pushing feathers through molasses. This is a trick of my mind. Predator and prey are moving at speeds close to fifty miles an hour, maybe more. They suck up the space without seeming to go anywhere, covering six, seven, eight hundred yards in less than twenty seconds. With intelligent precision, the deer turn and select the narrowest gap in the far tree line. The eagle closes in, moves to strike, but her wings tangle up in the trees and she lands, frustrated, and dangerously bereft.

Throughout the day numerous deer escape. The closest chance arrives in the late afternoon. Walking into a clearing, I come across a lone roe and stop still. She is close enough for me to study, has a precise, refined head, clear brown eyes and a black nose. She looks up over a crumpled mass of cover and faces me. Too close to be fair, trying not to draw attention, I remain silent. There is a dry click and snap of branches from behind as an eagle falconer moves forward. The deer bucks high up, flailing out of the low cover. Caught in a length of bramble, her grace is cut short, she panics and tumbles. Frustrated and fearful, she tries once again to rise vertically at full strength. Still caught, she twists, spins, landing on her back, slender legs and hooves pointing skywards. The eagle drifts over the top of her, folds and makes contact with a dull, concussive thud. The deer's scream is penetrating and loud, unpleasant, jarring and shocking. They begin to fight. In the mounting struggle the deer flips the eagle over, which then snatches from underneath, striking the deer's face, puncturing a hole. The deer rolls over in front of me, stands and runs right to left through the woods, passing in front of the other *berkutchi* and their eagles. One after the other the eagles are released. The deer flows and flashes between strips of thin pine, a majestic zoetrope momentarily frozen in movement, swerving and escaping each and every one.

Towards the end of the day the falconers slowly return from their respective hunts. We gather at the back of the community centre for a short ceremony to honour the lives taken. Pheasant, partridge and hare lie in the snow in front of the successful falconers. Two men step forward and begin blowing on brass trumpets and hunting horns. The sound is eerie. The music sharp then melodic, the notes smooth, erratic, piercing, playful and sad, like the movement of a deer or a hawk in flight. Local herbs and plants are brushed over the dead bodies. Some are butchered, shared or given as gifts; celebratory speeches and toasts are made. As I start to leave, two of the more experienced *berkutchi* invite me to join them for a more intimate and less formal gathering on their own land.

Germany

Early the next morning, after crossing into Germany I am standing in a hotel courtyard waiting and watching the steam swirl from a hot cup of coffee. I look down and distractedly crunch the top of the crusty snow with the heel of my boot, then flick the powder underneath with my toe. It reminds me of a crème brûlée. A pigeon suddenly smashes into the ground at my feet. I look at it; dizzied and stunned, it stands up, leaving a cartoon indent in eight inches of snow. Waddling away, flicking flakes from her feathers, it cocks its head daring me to laugh. Instinctively, I look up. The telegraph wire is rattling and vibrating. Ten feet behind the lines a large peregrine falcon, resplendent in full winter finery, silently chases a second pigeon across the hotel roof. She turns, sets herself in a low stoop and descends rapidly, raises her feet and snatches out. The pigeon rolls, then dips, sending the falcon wide, and drops down next to her friend on the ground. Noting my presence, and with what looks like silent malevolence, the peregrine crosses the winter sun, a scythe-shaped pitch-black profile, and glides smoothly away. Two trucks pull into the courtyard and the pigeons clatter from the floor back to the roof.

An hour or so later we pull up at an isolated enclosed forest area. The eagles are removed from their boxes and I watch in rapt silence as they are released from the falconers' gloves and allowed to fly free in unison. As we push through the coppice the eagles move above us from tree to tree, following the falconers. They land heavy and rough, curving thick branches down like the arm of a trebuchet. Large clumps of displaced snow slide and thump at my feet.

An eagle can easily scope prey over a mile away, decide to chase and leave the falconer a long way behind. They are also free to become jealous and potentially, to kill one another, or attack a human spectator. Free flying like this is highly unusual and risky, and teeters between control and chaos. A thin line between allowing the eagles to function as close as their natural counterparts while at the same time remaining biddable and safe. With a predator this big, I find the situation both scary and exhilarating. The relationship between these *berkutchi* and their eagles has been carefully orchestrated over decades of conditioning, knowledge and dedication.

I almost step on a hare. It lifts out from its camouflaged form, growing larger on long, lanky legs; it has big, bulging eyes and a russet coat. It opens up, sprinting out of the coppice and across the field. At the same moment, on the scent of the hare, a fox skulks over a low ridge. The dual movement prompts both eagles to heave out of the branches. Ignoring both the fox and hare, building speed, they instead fly from the wood to the outer edge of an old lake bed in the distance. A group of deer look up over a strip of maize. They start to run and dance balletic into the field. One eagle sweeps away from the herd and goes right, flying parallel between the deer and the trees, cleverly cutting off their escape route. The second female, following hard, gains ground and is almost on top of the slowest doe. The deer feigns left then jerks right. The eagle misses. The herd twist left and accelerate over flat, open fields. The attacking eagle lands abruptly; the other pulls off into the sky and makes a huge, sweeping return to a perch above my head.

Several seconds pass. I am unsure what she will do next.

From her lofty position she lifts off, cascading swirling drifts and droplets of snow on to my shoulders. I watch as she powers over the lip of a hill, flying low over the earth to a large drainage ditch several hundred yards away. Like a breeze block thrown through a pane of glass, she crashes into a wall of reeds, rears up, begins to flail, fight, and roll across the snow, holding the unseen fox. I clatter through the first set of trees and run. As I fall through ice and semi-frozen mud, my legs feel heavy, my muscles burn, the lactic acid builds painfully. The distance and conditions beat me. I am too late. I slow to a trotting pace, bend double, breathe heavily and collapse at a safe distance from the eagle. She is standing with wings spread, head crested in rage. The sweet, sickly stench of fox piss wafts around in the air. The snow under her talons is tossed up, and the tufts of fur, paw prints, droplets of blood and yellow spray tell me the fox has twisted out from under the eagle and skipped free into the drainage ditch.

The Hiebelers

The drive from Vienna airport has been long. It has also been more or less silent. An assistant of the Hiebelers, a man named Pan with a long ponytail and a thick beard, drives and speaks English about as well as I speak German. We gesture and smile but eventually find ourselves at a stalemate.

As we approach the lower Kamp valley the sides build in steep, angular granite, thick with deciduous woodland. The road thins and twist back on itself, following the contours of the river to my right. In the slick winter wet we seem to be skimming the surface of a large black snake. High up in the distance the sodium-orange glow of spotlights flicker through gaps in the trees. Stationed on the top of a mountainside a mile away, outlined by the orange lights, a gothic chateau hoves into view. Long, pale and majestic, several turrets jut up at either end. There are multiple balconies, arched windows and balustrades, all casting long shadows along the outside walls. I think of Dracula's castle, or perhaps the hideout of a James Bond villain.

At the top of the hill Pan pulls in through two wooden doors and comes to rest in a cobbled courtyard. Long past midnight I am led through a maze of passageways, up a staircase and on through cave-like corridors to my room. As I move to unpack, fake safety candles flicker with a weak yellow light. On the far wall I notice an opaque mural, a thin, broken, flaking image of a horse and a man. It could pass for a partial depiction of a hunting scene. I hop into bed and stretch out. Disturbed by my arrival, and through the darkness, the calling and *chupp*ing of a dozen golden eagles filters through an open window. There is something faintly ridiculous about the whole situation. It is like the start of a Hammer horror film.

For most of the following morning, I wander alone around the grounds of castle Rosenberg. It is even more impressive during the day. Walking across the cobbled courtyard, a small arched doorway leads to a huge rectangular lawn surrounded by neatly trimmed boxwood hedges and gravel paths. Behind and above me is a long passage with thirty curved open-air arches, like the base support of an aqueduct. I walk across the lawn and peer over the wall; the smooth outer side drops down a hundred feet or more before touching bumpy rock. The woodland below stretches out as far as my eyes can see.

Back inside the castle walls, in the far-left corner is a crèche. Four fledging eagle chicks are sitting on the edge of a man-made nest. One pushes her leg out, has a stretch and yawns. At around thirty-five days old, they are already the size of a cocker spaniel. Another five or six adult eagles are dotted around the grounds on block perches, or on high perches strung between trees. In an outer atrium two vultures spread their wings six, eight, ten feet wide, and yaffle at me through the mesh screen of their aviary.

Josef and Monika arrive and greet me. Josef is short, compact and powerfully built, a true patriarch. Monica, matriarchal, is no less an impressive physical presence. They speak no English, but we begin to communicate and muddle along well enough. It is clear they both have a lot of work to do, are focused, efficient but not unfriendly. I follow them like a lost child.

I am taken through the main corridors and enter the central body of the castle. This area has been converted into a small museum containing stuffed birds, paintings and cabinets full of gloves, leashes, bowls, knives, hoods, feathers and rare, historically significant falconry treasures collected or given to the Hiebelers as gifts from around the world.

Josef's experience is unquestionable. He travelled to the Russian steppes to study with the Kazakhstani *berkutchi*. He has worked with the university of Almaty (Kazakhstan), the zoological institute of Bishkek (Kyrgyzstan) and the University of Heidelberg (Germany) on eagle-breeding projects. The eagles Josef and Monika breed and own are some of the finest found anywhere in the world. With his experience of the Kazakhstani *berkut*, the relationship he has with his eagles is an unbroken lineage between the past and present, between ancient East and modern West. Josef and Monika's understanding of eagles corresponds directly to my experience with the goshawks in Pakistan. We have shared witnessing traditional falconry first-hand and are connected tangentially by our different journeys to the East.

As I observed at the field meets in Germany and Slovakia, golden eagles need to be muscle-fit and motivated. If they are not, they will fail. If they fail, they will lose confidence, become lacklustre, frustrated, aggressive and difficult to handle. An eagle has a lot of muscle to build and a lot of dangerous aggression to direct. The landscape of Kazakhstan is one of mountainous ranges, sheer cliff edges and steep valleys. The *berkutchi* use gravity and long distances to help build the fitness and muscles of their eagles. In Austria, under the auspices of Josef and Monika, this traditional vertical training has taken on a distinctly modern twist.

We move from the museum back out to the grounds of the castle. Of the dozen or so eagles on display, two particular juveniles have waited patiently for their training session to begin. Monika, Pan and a female assistant climb into a people-carrier and drive down into the valley below the castle, while Josef and another female assistant remove the two eagles from the blocks. We move quickly, turn right through a passage then across the courtyard and down into the belly of the castle. The bulk of the eagles throws looming shadows across the curved, cream-coloured ceiling; their excited *chupp*ing echoes across the stone walls and floors as we arrive on a private balcony overlooking the valley.

Below us, the river appears as a silver strip. The red-topped slates of the houses are tiny, like in a medieval model village. To the left of the ribbon of water is a small patch of grass, the size and shape of a lime-green face flannel. When Monika and Pan arrive the people-carrier appears as small as a Dinky toy. A white speck emerges from the trees at the side of the field, a Lipizzaner horse ridden by the female assistant. Josef makes a phone call and the horse starts to gallop from one end of the field to the other. Fastened to a rope and being pulled over the grass is a whole, skinned deer carcass. Josef removes the hood from his eagle and propels her forward out into the sky. His assistant does likewise, and the smaller male follows. Both eagles power down for a few hundred feet, Josef's female the more determined, folding her wings and stooping with the velocity of a falcon. The male, sensing he will lose and naturally subservient, takes a more leisurely route, circling more slowly, his wings spread wide as he begins soaring, turning and curving above his sister. She reaches the deer carcass at full speed, lifting it from the ground, and the horse pulls to a stop just as the male eagle arrives.

I am invited to swap positions and the training session begins again. This time the male is much faster and reaches the lure first. Once the vertical training is complete, for the next hour the eagles are flown repeatedly from the glove at the horse-dragged lure. Often the eagles arrive together and begin a sibling crabbing squabble. The eagles have no preference or gender bias, just the desire to be handled correctly. Picking up on their argumentative and negative behaviour, Josef and Monika move and react as one, distracting and diverting the aggression of their eagles with hand-held rewards. Rather than the situation building to a dangerous confrontation, with remarkable softness the eagles take turns feeding on the large pieces of proffered deer flesh. The repetition and precise handling is enough to displace latent belligerence, cleverly condition tolerance and build muscle fitness. The whole process is an impressive orchestration of time and effort. The context and setting are the result of dedication to the lives of golden eagles over several decades, and the techniques a window into a 5,000-year-old past, that is still appropriate, relevant and workable in the present.

The Hiebelers, along with several other key speakers, would make a presentation in the seminar tent I organized for the first UNESCO festival of falconry.

The Festival

The first international festival of falconry was held in the grounds of the Englefield Estate in Reading. Hundreds of falconers from around the world attended. In and around the site, traditional tents and encampments of the various International Association of Falconry members were set up. People of various languages, colours, creeds, dialects, cultures, races and religion intermingled in a glorious celebration. Specially prepared foods and drinks were brewed or cooked over open charcoal fires. The variety and scope of the dress worn by the human participants and the birds of prey on display were magnificent. There were stalls, education/seminar tents, falconry displays and a parade of the nations. The organizers created a Glastonbury for falconers. As the evening arrived, the inevitable celebrations escalated and gave way to the late-night madness often found on Worthy Farm.

The history of English falconry has always been one of class. Since the very beginning, there has been an uneasy truce between landed gentry, royalty, their long-winged falcons and the lower-level *austringers* with their goshawks and sparrow-hawks. In his superb book *The Hound and the Hawk, The Art of Medieval Hunting*, the historian John Cummins quotes a description from a medieval falconry treatise:

> *When one sees an ill-formed man, with great big feet and long shapeless shanks, built like a trestle, hump shouldered and skew backed, and one wants to mock him, one says, 'Look, what a goshawker!' I know the goshawker would like to beat me for this, but there are two dozen of us falconers to one of them, so I have no fear. Goshawkers are cursed in SCRIPTURE.*

The echoes of this ingrained prejudice still exist. In the evening there was a huge medieval-style banquet of roast quail and vegetables, followed by a ceilidh. Tickets were limited and pricey, and only those on the list were allowed access. As the evening meal ended and the music and dancing began inside the tent, the lower orders, and those without tickets, quite wonderfully proved that times change very little. Bottles of traditional spirits from the Balkans and further east were swapped and mixed with moonshine from the States. Not to be outdone, the Irish and the Scots contributed their own whiskies to the testosterone-fuelled antics.

Running with both the fox and the hounds, I left the ceilidh and joined the bacchanalian excesses outside the banqueting tent and around the festival grounds. Songs were sung, instruments played, and falconers boasted who had the best hawks and the best bloodlines for breeding. Thick-set Americans and bawdy English goshawkers wrestled one another to the floor, ripping clothes and breaking bones; an ambulance was called. Fires were lit, friendships made and lost in minutes. A quad bike belonging to the chief organizer was 'accidentally borrowed' by a drunken heathen and driven at high speed into a pond. On the large lake, ornamental domesticated ducks were hunted with a catapult, thankfully without accuracy; none made it on to the barbecue.

Merely spectating and avoiding much of these excesses I took the opportunity to talk to an American falconer. Craig told me of his adventures trapping migratory peregrines in South America, the historical background to Japanese goshawking; how they hunt golden pheasant in the emerald bamboo forests with pure white goshawks. I shared my stories of Pakistan and Europe. A friendship was formed and he invited me to visit Illinois and drive across country to the plains of North and South Dakota to trap a prairie falcon and fly gyr falcons. For ten days over Christmas and New Year the following year, that's exactly what I did.

Illinois

Eagles and hawks are usually flown from the glove in direct pursuit of fleeing quarry. It is a flight style that runs parallel to the land. Falconers (as opposed to *berkutchi* and *austringers*) are those who fly peregrines and other long-winged falcons. They exploit the natural proclivity of their raptors to fly high and stoop at quarry on a vertical axis. To perfect the flight style, to enhance pitch and power, they train them in a specific way. Northern Illinois in winter is cold, flat and wide, perfect for training falcons.

Driving across the state with Craig, we meet up with his friend Frank. As we step from the truck on to the training ground and prepare the falcons, the wind is a roaring, sharp white noise. Watery tears are blown sideways and I can feel them begin to crystallize at the entrance of my ears. My ungloved hands and face are burnt frozen in seconds. I watch closely as Frank removes the hood from his falcon. This particular falcon is big, well over two pounds in weight, with rare, highly prized, dark feathers. Poised in contrast against the snow, the falcon appears as negative space, a charcoal void shaped like the flat edge of an oval spear. Unconcerned with the conditions, he rouses, trapping fresh air near his body, turns his head skywards and launches off Frank's glove. His pointed wings are bigger than a peregrine's and move in wide arcs, tip to tip, high above the shoulder then deep below the chest.

In tight circles, he rides through the wind and climbs as if suspended in a colourless tube. Mounting into the sky, spiralling up, he deviates no more than thirty to forty feet either side of our position. At around 800 feet and directly overhead, he looks like a ladybird climbing to the top of a distant windowpane, a slight speck and fleck in the sky. Before he spots something in the distance and begins to drift away, or lower his magnificent pitch, he needs to be rewarded. Frank removes a live duck from his bag, looks up, shouts and lets it go. Startled, it falters, wobbles, and takes flight. The falcon drops with laser-line accuracy, passing through the duck 50 feet above the ground, splitting flesh and breaking bones. The falcon follows the dead duck to the ground. Frank cuts the quarry open, the falcon's reward of warm blood and fresh flesh smearing across his chest. It looks simple. To me it looked perfect, a textbook training session. Frank has been trapping, breeding and flying these big falcons for nearly forty years.

He turns back to the truck, having noticed slight faults, and mutters, 'It could have been better. He needs to be better for sage grouse.'

South Dakota

Later in the week we drive with Frank and his falcons across the US to a hunting lodge shared by friends. We take our time, trying in vain to spot and trap a wild gyr. We find one near a town called Pierre. She is too old, what falconers call *haggard*. We watch her high on a pylon near a lake and decide to leave her alone. As we cross the state lines, the conditions are all wrong. We stop to talk to a farmer and ask if he has seen any falcons. He has not: "Too clear, too windy."

Craig and Frank continue talking to the farmer while I stare through the window into the back of his truck. A bright reflected light bounces across the scuffed aluminium base and blinds me to almost everything but the dead coyote. Circles of snow lie unmelted across her forehead. Facing the sky, her eye is small, shrunken, a milky, frost-coated marble. I get out, take pictures and touch her. She has a soft fur face with muscles as hard as steel underneath. A rictus snarl spreads across the muzzle. In minus 35°C, the Alsatian-sized animal is frozen solid.

The land behind the coyote rolls back to a thin, brittle

panorama, an undulating space bouncing subtle blues, zinc, titanium whites, copper-green and silver patterns within lattice-work ice sparkles. It is a harsh kaleidoscopic desert formed in cold tones.

Visually, a desert of snow and ice appear as the elemental opposite of the deserts of sun, fire and sand. The extreme ends of cold and heat do, however, create similar environmental conditions. Both deserts are dry, clean, and free from bacterial diseases. A lack of running water makes dehydration easy. Both are wide, distant and unforgiving. Both have delicate, intense seasons of life and death. For these specific reasons, both environments produce equivalent birds of prey.

Gyrs are the largest and most powerful falcon in the world and are highly sought after by falconers from around the globe. They range nomadically across Arctic tundra, stark frozen seas, from Greenland down through Alaska into Middle America, and finally to northern Europe. Despite the specifics of these cold conditions, the gyr falcons of the northwestern hemisphere share the same biological characteristics as the hot-desert falcons of the Arabian peninsulas. The American gyr falcons take on a different form but have evolved in parallel with the sakers, luggers and lanners Salman rehabilitates in Pakistan. These two very separate species, both from markedly different environments, have more in common than they do with any other falcon. The same can be said for the falconers flying them.

We pull away from the farmer and keep driving as dark descends. The shared lodge we are heading to is close to the wintering grounds of the most significant indigenous game-bird on the plains of North America. Sage grouse have been here a long time, long before modern America was invented. They feature in ceremonial dances of the Lakota Indians and were once as plentiful as buffalo. Their numbers have dwindled due to modern farming methods and habitat loss, but they endure in numbers where ranchers are careful, and where falconers and gun hunters help preserve natural wild spaces. Highly sensitive, sage grouse take flight at the slightest provocation, at downwind speeds of over 70 miles an hour, they cover huge distances in a matter of seconds. They are tough, will withstand a direct hit by a falcon and despite being driven hard into the ground, will bounce back up unscathed and escape. Equivalent to the goshawks, francolin and quail of Pakistan and Croatia, or the eagle and deer of Germany, the gyr falcon and the sage grouse are a perfect match.

I wake in the lodge early and stare out the window. The wind is shrill in the blackness and howls all around the lodge roof. Deep and snuggly warm in a wooden bed, the thought of getting up is difficult. By the time I am ready, the falcons and dogs have been loaded into vehicles. We trundle slowly out into the snow. In the back of the truck I rest my forehead on the cold window. My breath condenses, omelette shaped, then evaporates. Through the glass, the sky appears as a water-colour band of pale green spread below dark Prussian blue. The clouds begin to stretch and pull apart and the sun clips the edges, turning them into candyfloss-coloured drifts a mile long. They curl across the curve of the earth like soft, pink Persian slippers. As we approach the feeding grounds of the sage grouse, the multicoloured dawn slowly morphs to a shade of steel. I see a brief shadow and a slight movement behind a bone-grey post. A lone grouse shuttles along from under a bush and takes flight, leaving a light puff of snow. He rises directly into a stern crosswind, banks sideways and flies a mile away in less than ten seconds. Frank and Craig note the movement. I am told to look left. The truck rocks, buffeted sideways by wind, and the metal edges of my binoculars bang the glass. In the distance, at ease and thrilling in the conditions, a large flock of pigeons twists in bumping swirls, fragments of black ribbon spinning in coils high in the sky. I turn my gaze elsewhere. Both falconers laugh and tell me the pigeons are sage grouse. When I try to refocus, the covey has gone. Seemingly impossible, exquisitely unobtainable, they are the fastest, highest-flying gamebird I have ever seen.

It takes an hour to track them down, and we find them crouched and hidden in a wide, flat field. Using the cover of a dam wall, we get into position. Standing on the highest point, the jerkin's (a male gyr falcon) hood is removed. He rouses his feathers and is away into the sky. An elite athlete sprinting up a glass staircase: three, four, five hundred feet, and climbing. Frank runs into the field. Twisting kite-like on invisible string, the jerkin continues to circle, rising higher in a tight corkscrew: six, seven, eight hundred feet, and beyond. Frank is halfway out and his dog is even further. Two grouse break early, three others lift up and curve downwind. The gyr oscillates, hesitates; the rest of the covey rise en masse. Twenty or thirty grouse cut and twist in different directions, a perfect bamboozle. The gyr, just out of position, falters, selects the slowest, sets his wings and folds over in a steep, parabolic arc, a pendulum gaining momentum. At a hundred feet above the ground the jerkin levels out behind the grouse and they fly horizontally, locked together for over half a mile, a marathon into the wind. With compact, blithe power the grouse rises into the sky, taunting, and confident. Sensing weakness in the young gyr, it begins to accelerate. Frustrated, the falcon digs deep and opens up, the depth and pace of his wings now fully formed. Just at striking distance, the unconcerned grouse subtly twists, mounting even higher over the jerkin's head. This is too much. He slows, pulls off and circles around in a long left-hand turn. The grouse disappears into the horizon.

It takes a minute or more of hard flying before the jerkin is back above Frank in the field. Frank is walking slowly towards the dam and the disappointment felt by the members of our group is tangible. The jerkin, still dwarfed by the skyline, remains high, floating about in a tight figure-of-eight loop. As Frank reaches a ditch below the dam wall he startles a hen pheasant into flight. The jerkin, slightly offset from vertical, drops in slick silence; a black-lead spear, it is breathtakingly beautiful. Three, four, five... long... slow... seconds; a huge stoop reaching 80 or 90 miles per hour. The pheasant is tissue thin in the face of such force and crumples. The impact is audible, like a lump of wood hitting wet carpet. On the ground and with focused aggression, the jerkin repeatedly stamps the dead pheasant into the snow. The final blow cantilevers the carcass across stubble, and buds of blood and frozen earth scatter in the wind.

As we drive back to the lodge, the wind has pushed the cloud formations away and the sky is a light blue, with a pale, watery sun sitting low on the horizon. Crossing a small creek, we spot a pair of mallard ducks tucked down and protected in the curl of current. Still early morning and with another falcon to fly, we decide to take a chance. Removed from the truck, the falcon rouses her feathers in preparation for flight, then squirts a thin mute (faeces) on to the snow. A youngster and inexperienced, as she lifts from the glove she is hit by a crosswind and topples unceremoniously to the ground. Regaining composure, she sets off out over the fields. Adjusting to the bowling, blowing pressure, she begins to fly slowly into the surging wind, climbing high over us. Higher. Higher still. To the uppermost reaches required by a trained falcon. Close to a thousand feet. Any higher, and she would disappear from sight. We walk back down the road, hop over the fence and run at the stream. Unseen before, at least fifteen ducks take flight in different directions. A single mallard flies away then mysteriously doubles back, coming directly towards me. The falcon is in a pure vertical descent, a perfect teardrop stoop. I turn back to watch the duck. On the periphery of my vision a black, threading line, a thin, blurred streak, rips across its neck. The duck folds over and hits the ground with a muffled *whump*. Feathers from the impact spin on spines and skid across the snow, like miniature sailboats adrift on a sea of crystal white. The duck, coated in thick winter wrapping, is only stunned. Standing, then flying to sanctuary, she slides into the water. In my inexperience I rush to re-flush before the falcon has regained significant height. The falcon lunges, the duck dives, the falcon becomes soaked and useless. Popping up downstream, the duck whirls around in the current, faces me and quacks.

Chicago

On Christmas Day, I drive with Craig and his friends to the edge of an out-of-town shopping centre surrounded by food outlets, roads, housing, traffic, trains, work spaces and car parks. These buildings have been built on the feeding grounds and nesting sites of migratory duck and geese. The whole area is block upon block of governmental buildings; Lego offices, uniform, blank. Dead snow in a dead land; brown litter is blown about and pushes into bricked corners. In between the sprawling mass of low-rise, boxed-off offices, several neat man-made ponds have been dug to soften the edges of the angular architecture. Nearby, a stream fills the ponds and flows behind an industrial estate and a cinema. Beneath the water, cloth and discarded plastic billow in the current. Trapped between rocks are a wallet and a single shoe. A purple grey scum coats the roots of reeds. The ducks slot in, paddle and puddle about in shallow, acidic water. We prepare the falcon in the car park of a supermarket. A police car pulls up in an alleyway. The officer stares, we wave and he drives away. The falcon, set free and flying, disappears. A minute later she reappears over the concrete edge of one of the buildings, no more than a hundred feet in the air. We flush the ducks into the contained and controlled space. They cannot move or escape, and one strikes me on the shoulder, knocking itself unfairly off balance. The falcon hits it easily. Thankfully, the duck survives.

When the New Year is out of the way, my final few days are spent hunting rabbits in the woods with a goshawk. Craig smashes his ankle and is in pain, so we stop and I prepare for home. It was a strange and flat ending to my odyssey across America but, somehow, it was fitting.

It was the last trip I would take for over a decade.

Endings

What was it that attracted me to the different people I met, to the places I went? On the surface, it is obvious – falconry. It is true that, initially, my motivation was to learn and experience as much as I could about the eagles, hawks and falcons, their quarry and the landscape. As the next few years passed a gradual, more subtle set of values and observations became apparent. It was then that the behaviours and personalities of the falconers became as significant as the birds of prey themselves.

From scientists and atheists to the devoutly religious, Christian to Muslim, pale pinks and whites to light brown, walnut and black, short to tall, well educated to illiterate, male and female, adults and children, rich and poor, the falconers I met on my journeys were the most experienced in the world, and as varied a cross-section of humanity as anyone is likely to find. Each transcended the confines of these boundaries, relying instead on something far larger than limited man-made definitions to create their identities. Each shared a global commonality, a collective, overlapping character driven by the only thing that mattered to them: constructing a balanced working relationship with wild birds of prey. In questing for and creating an exacting feedback loop of mutual trust between themselves, birds of prey and the natural world, they excelled in the best qualities we own as creatures. In their element, rooted in the specifics of their landscapes, all were explicitly conscious of what it means to live with dignity and freedom. Without exception, they were kind, generous in thought and deed, lacked greed, were communicative, humorous, relaxed, open and confident. They all exuded self-sufficiency and carried themselves with sincerity and honesty. These characteristics were most evident in the tribal falconers but were threaded through every other culture from East to West. Falconry for them was not simply a way to pass time or a means to distract them from the vicissitudes of living. Birds of prey were the reason for life itself, nourishing them directly with food, with breeding projects, in university papers and PhDs, in work and in self-contained psychological peace. This level of commitment and connection went beyond any conception I had about falconry. At the time of each trip my relationship with the natural world, and hawks in particular, remained mostly separate from my lived life. Although I was passionate and enthusiastic about birds of prey, falconry was a bolt-on accessory in the form of holidays or one-off adventures, a quaint hobby, something I did in my spare time, outside of teaching. I did not define myself as a falconer, simply because, in comparison to the broad spectrum of people I met when travelling, I was merely a pretender, just skimming the surface. I was a tourist, a passive spectator, at best.

This was about to change.

3

The Drop

Tomorrow afternoon will be another riotous summer day. After their last free meal I will walk the sparrowhawks from my cottage across the fields to a large privately owned stately home. Behind the mansion is a long, shallow lake, surrounded, shrouded and protected by a secret covert of oak, larch, ash, beech and pine. It remains untouched, is overgrown and devoid of overt human interference – bursting with natural life, food, water and shelter. I have no doubt these sparrowhawks will survive when set free.

Every hawk I have known or owned has been unique. These rehabilitated sparrowhawks have been no different. Each hawk is hard-wired to act and react in their own particular manner. Each displays instinctive quirks and differences by degrees of evolved behaviour. Throughout their training and rehabilitation it has been possible to tinker at the edges, manipulate and nurture certain aspects of these characteristics, but once fully grown, each hawk varied little in its specific reactions to the world. I can no more change the innate characteristics of an adult sparrowhawk than I can my own.

The writer Ben Okri describes being born as 'a shock from which I never recovered'. I concur. I took a long time to be born and came out silent and angry. By all accounts, I glared at the doctor, pushed and kicked, then urinated over him. I have struggled with change and with my reactions to the human world ever since. I had my first counsellor when I was fifteen and a wide range of different therapists have passed through my life ever since. Their methods (although enjoyable and interesting) presupposed that the hard-wiring underpinning my thoughts, reactions and level of engagement were normal. It was not until I was around the age of forty-two that a clever, sympathetic and highly perceptive analyst suggested a reason for my sense of dislocation and difference, that the primary motivating impulses for my oddities and behaviour ran a lot deeper than I had appreciated.

'Asperger's' has a whispered, hissing sound and is slimy. 'Autistic' sounds like 'spastic'; it feels spiky and unyielding. 'Spectrum', on the other hand, is taut and twangs, a natural refraction of pure white light split open to reveal interwoven changes of colour. The phrase 'autistic spectrum' describes a rainbow range of odd thoughts, feelings and behaviours, a series of strange motivations arching wide and bright across the human mind. Each recipient is different and unique; some are locked away within repetition and precision, others are highly functional, seemingly floating through the world without too many issues. The only shared feature from either end is that we experience the world very differently from a large proportion of the rest of society.

On the outside, I appear normal. In my head, I feel like a badly levelled graphic equalizer. I have spent a long time covering this up, muddling along and pretending.

For as long as I can remember I have experienced the world erratically, as if aspects of my development have been arrested, or pitched on widely different levels. The overarching sensation I have daily is one of general confusion, fear and anxiety.

On a one-to-one basis, unless I know and trust an individual well, I lack the ability to consistently judge what people mean in conversations. I find it difficult to read facial expressions, intonations and meaning. I constantly strive to find clarity in the words and actions of others. I experience a basic conversation as multiple strands of meaning and trying to understand each unit in a logical order often overwhelms me. My mind panics, is slippery, steps ahead and jumps about with three or four alternative interpretations running concurrently. Rarely do I hear a full conversation; they exist predominantly in fragments and snippets of language. I constantly look for a speedy way out of contact with other people.

Hypervisual, I have no filter and get distracted by the trivial, say, a hair on the corner of a shirt, I jiggle my legs about and avoid eye contact in order to concentrate. In extreme moments (when people step too close or arrive unexpectedly) I have the sensation of being tiny; the whole world feels amplified. A raised eyebrow is a tidal wave and can throw me into a confusion of interpretation. It feels as if I am living on the surface of the other person's face. When I am at my worst, I can spend several days replaying the words of a five-minute conversation, trying to decipher what was said and, more importantly, what was meant. Of course, within that timeframe I have had other conversations and so the layers of conversations and thoughts build and build in a messy overlap.

However, I have no problem discussing things that interest me. I can happily stand in front of a hundred people talking about hawks, or before a whole class of children and discuss the merits of a painting. As long as there is a conduit through which to discuss a topic I am interested in, then I am fine. Put me around a dining table with three strangers, though, and I fall apart. To stay on track, I often use pre-developed routines and stories prepared days in advance, and for a large portion of the time I exist in jokes; laughter in other people is an emotion I find easy to gauge. This makes me appear, superficially, charming, confident and free-thinking. In fact, I am trying to control the situation to gain the type of attention I can judge, to short-circuit my general confusion and anxiety. Consequently, I have a vague grasp of social conventions but no concrete understanding of hierarchies. They do not seem to make sense. I will blurt out the most inappropriate things without concern and with little realization of their effect on other people.

My emotions often arrive at the extreme ends of expression: black or white, good or bad, right or wrong. I either understand something or dismiss it completely, like a petulant child casting aside a particularly complicated puzzle. I find it difficult to organize coherent abstract emotions so, unless I have experienced a situation directly, I have a limited degree of empathy and can easily disengage from complex emotional situations. I get bored, abruptly break away from friendships, extended family and other relationships, preferring the safety and relative stability of my own company.

Without a specific location to visit, a set task or a pre-considered activity, walking through a town often feels illogical, frightening, complicated, synthetic and jarringly unnatural. Large crowds and shops, supermarkets or city centres often cause me distress. I genuinely do not know how we have ended up with the world we have created. Or, rather, I do know; I am just at a loss as to why.

Left to my own devices, I fall into self-made routines and find it difficult to compromise. Unexpected change or the potential for rapid change becomes a serious problem. I prevaricate, almost frozen to the spot with fear. When under stress, not knowing which way to turn or what I should do, I react badly. In preparation for fight or flight, trying to find a pattern, to predict an outcome, I make rash choices and stick rigidly to them. In these moments, I have no deep comprehension of my actions and can appear amoral and stubborn.

Even when not under stress, I feel a constant, restless humming under my skin. The wheels of my mind spin wildly, a self-perpetuation machine of continuous thoughts, fragments of memory, looped stories, images, sounds and details of conversations had that day, or from twenty years ago. I experience a constant wall of internal sound, a million different screens playing oddments and snatched moments of my life. When the noise becomes too much, I move. I move a lot. Inertia blinds me. I cannot sit still for long. So I walk, usually the same route, engaging in long, meandering conversations with myself, internally and externally.

Possessed with an acute level of hypersensitivity, I find it easy to tune into external movement and seem to notice and remember things of interest with an alarming level of clarity.

Nature changes pace slowly and is therefore relaxing and intimate: quiet clouds moving, a change in the spread of light
over a landscape, a shoal of fish, a leaf falling to the floor, a bird
flying, a wiggle of a tadpole's tail, mayflies hatching. I can stare
at a stream, mesmerized, for long periods of time, hypnotized
with an almost synaesthetic sensation. I have an affinity with
and understand these things because they are gentle and clear
and reciprocate my existence on a level I comprehend.

If ever there was a state of mind or a set of behavioural patterns guaranteeing failure in close human relationships but
success with the natural world, then mine is it.

In 2010, at the age of thirty-six, I had a baby with my long-term
partner. My son took an extraordinarily long time to be born.
His mother is tough and, after nearly ninety hours of labour,
he finally arrived in the world. I was not prepared for the vast
confusion a birth precipitates. The experience was violently
overwhelming – deeply complicated. Too many people
became involved – family, strangers and medical staff. There
was too much noise, too many internal voices, self-invented
scenarios and conflicting destructive thoughts zipping and
zagging within my mind.

After a few hours my son turned yellow and became jaun-
diced. He was placed in an incubator and I looked at him
from a distance. Bathed in ultraviolet light under smooth,

cast plastic casing, he was an odd little chap, with tiny nails, pink, flaky skin and a grumpy face. As more and more people became involved, I felt very much out of control. Terrified and marginalized, I was suddenly a minor character in a major life event and felt forced into a role I did not understand; I felt as if I were falling headlong into an abyss. I had no time to decompress, to take it all in and to work out the meaning. I wanted everyone to leave him alone. To leave us alone. It never happened. I had no time to bond. Any love that may have been present was squashed. Overshadowed by the vertiginous fear and complex inter-family dynamics that a birth triggers – my son was a Pandora's box built out of wrinkled pink-and-white human flesh.

On his arrival home I kept looking at him. He was certainly interesting, and he smelt wonderful. He just did not seem to belong to me. He was nothing to do with me. I knew this to be terribly wrong. I searched endlessly for some positive emotion and sincerely struggled to find a discernible connection. I never experienced the famed rush of love common to those who have children. Quite the opposite. The only tangible emotion was a rushing urge to get away. The guilt and sheer magnitude of this void, the weight of not having a normal reaction or range of emotions, was petrifying. Every time I picked him up I felt deeply inadequate. I felt dead inside. I had absolutely no idea how to be a father. As the days and weeks passed, the weight of this mounting dislocation grew. Men never mention having a lack of feelings towards their children to this extent, and I had no outlet to express how I felt. I did not dare. In comparison to the emotions expressed by the rest of my son's extended family, I felt abnormal, inhuman almost sociopathic.

Unable to express anything other than fear and shame, I shut down, mentally locked myself away, became indifferent, resentful and angry. I hated myself and the situation with a rage and fear beyond safety. I was not a fit human, let alone a parent. I didn't even want to be a parent and said as much to my partner. Over several months, my behaviour, reactions and conduct were understandably too much for his mother and she eventually took her son and left.

Within a few weeks of their departure, and without warning, I was handed a redundancy notice. The school I taught in had been earmarked for closure. I had been there for over a decade, had been successful and achieved consistently outstanding results. The job – a vocation – was creative and free. It was who I was for twelve years. I defined myself by my job. The forced removal of this locus, with its set routines and success, pitched me into an even steeper sense of crisis. Without pause, on autopilot, I tried to do what was right. I moved from a place I loved to the southern edge of England, to be near my son and start a new job.

It was an unmitigated disaster.

In a vain attempt to re-create the best bits of my previous

life, I rented a house at extreme expense to be near the countryside. To anyone of a normal mindset, a country lane is a country lane, one field similar to another. But landscape is not generic for me. Location and environment are specific and particular, the comfort and security arriving in the shape of resident animals, plants and geography. In my previous home I had a clear mental and physical picture of my location. I knew where everything was, where to find things, where things grew and under what rocks to find particular animals. In this place I only saw a simulacrum of a view; there was slippage, it was uncanny. At a distance, it seemed welcoming and unspoilt; up close, a different version presented itself. I walked the lanes and fields trying to connect, to find some form of recognition. Instead I found a dead cat in a black plastic bag, fur falling and peeling from bruised skin. No bigger than a finger, her soggy newborn kittens were suffocated beneath her. I found litter and endless fly-tipping. I found old, rust-coloured tampons on paths. I found discarded fridges with black meat bloated and floating in yellow fluids. I found stained mattresses, dead ponds, dirty vibrators and discarded condoms. I saw dog shit in bags on posts or thrown into trees. I heard nothing except the hissing roar of motorways, saw nothing but extending, encroaching buildings, burnt-out vehicles, barbed wire and no-entry signs. Instead of cycling to work, my commute was long, dull, pushing along tarmac. At the end of the road, my new job. The antithesis of what I knew education to be.

Owned by a corporate conglomerate, it was a business. The
art department redesigned for the modern world, open plan,
an office of glass, steel, strip lights and smooth walls. I followed
strict units of work, implemented other people's ideas, producing forced, confined and prescriptive art. The chaotic,
hilarious, beautiful children were the only respite from a job
that was just a job. The school was built simply to contain and
condition pupils and teachers alike.

There was no discernible space or time for me to accom-
modate or understand these very serious changes. Struggling
with the complexities of having a son, grieving for the loss of
my previous job and 300 miles from home, I was physically
isolated and floating alone inside my mind. I saw no way out,
no way forward and certainly no way back. Frozen in panic,
I carried around a relentless, raw anxiety, manifest in a self-
destructive dysfunctional anger. I was deeply unstable, fractured, adrift and helplessly lost. Not wanting to stay, unable
to leave, I slipped into a weird liminal space between being
present and being disturbingly detached. Any spare time I
had was spent aimlessly driving around the south coast, then
up and down England, occasionally sleeping in the car at
remote, isolated spots.

From this fragile half-existence I ran towards anyone that
looked remotely stable and normal. The woman I met was
an immediate emotional and physical release. Within a few
weeks, I literally ran away. Leaving a lot of my belongings in

the house I rented, walking out of my job, I turned my back on the whole situation and married her.

It was insane.

The relationship, built precariously on the shifting sands of my injured psyche, constructed for all the wrong reasons, was never going to last. Within eighteen months the initial fun and momentum had gone, superficial charm had been exchanged for reality and the cracks appeared. On the final day she turned up with a white van to collect her belongings, so I left. On my return, everything had gone. Water poured out of pipes across the kitchen where the washing machine had been removed. Where the carpets had been glued, patches of grey foam underlay remained on vile yellow linoleum and bare patches of concrete. Electricity about to run out, the phone line due to be cut, going by any standard measure of success, the résumé of my life was total failure: a career, a family and a marriage all burned out, done and dusted within four years. I was forty, unemployed, penniless and about to become homeless. I had not escaped. My choices and conduct had merely compounded and concentrated the pain. I was in a space and place of my own making, I was unknown to myself and utterly exhausted. Statistically, the biggest killer of men between the ages of twenty-five and forty is suicide. I sincerely felt its draw, felt the attraction of oblivion and of blank freedom far beyond the situation I was in.

~

Before my son was born I lived in a small house in the grounds of a dilapidated stately home. It was here that I brought my son home from hospital. It was his first home. It was where I trained Cody and witnessed the sparrowhawk chasing a blackbird through the garden. Slowly, the surrounding coach houses, stables and, finally, the manor house itself were developed into an exclusive gated community. Over the course of construction I befriended one of the bricklayers who lived on site. He owned a small 300-year-old tumble-down cottage nestled in the nearby woods. I had seen it on numerous occasions when walking and felt an instant affinity with it. It reminded me of the cottage I had lived in as a child, and I had always wanted to live there. Throughout my travels and the birth of my son, I remained intermittently in contact with the owner. In a last-ditch attempt at finding a home, I emailed him. His reply was succinct, two lines providing a doorway, a small grain of hope and a literal lifeline. The cottage had been empty for six months; I could move in if I wanted. So I ran again, only this time on my own, to a place of safety and solace, carrying a small amount of cash, art materials, a fishing rod, a bed, books and a dog.

~

The Cottage

My earliest memory is from around the age of two. I am holding a crayon over a picture of a fire engine and the sun is shining through a window on my right. My first original work of art is a viridian-green chameleon with blue claws painted at the age of three. I know this because it still hangs on the wall of my father's home. I have been a visual artist for as long as I can remember. Painting is good therapy. It requires a focused routine. You have to relax and let yourself go. The process connects to a vast inner space, a peaceful and powerful place where the fuzz and pop of a busy mind stops. Travelling to it through brushstrokes and pencil marks, as an image builds, the feeling of release and sense of self-worth art gives me has remained consistent ever since I pushed a black crayon across a fire engine in the sunshine.

At the cottage I painted what I saw for up to sixteen hours a day, seven days a week. Orange-and-black great-crested newts, horses, leverets and hares, lapwings, sparrowhawks, pigs, pheasants, bats and birds. As the work began to build, I hung everything on the outside walls of the cottage and made two signs. I hammered them into the roadsides at the top and bottom of the lane and began fishing for commissions. On the first Sunday I sold a large drawing of a goshawk. The following week it was a picture of a dog. With the money I built a portable exhibition space and took my work to country fêtes, horse events and agricultural shows. What started as a way to clear my mind, a way to deal with failure, slowly became a living. As the summer rolled around I supplemented my painting with seasonal ground work and gardening on private estates, or for anyone who needed help, and became self-employed; never again would I be made redundant, but I would remain poor.

There is no romance in poverty. Or indeed in hunger. Context makes a marked difference to the degree of fear any individual experiences when confronting both. I have no idea what a single mother with little education trapped in an inner-city flat feels like when facing her own personal wasteland of distress. I know homeless heroin addicts die young. I will never feel their subtle decline or abject despondency. I was not John Healy in *The Grass Arena*, nor George Orwell on *The Road to Wigan Pier.* But in the overlap between arriving at the cottage and becoming self-employed I experienced the shame and humbling irony of sitting next to my ex-pupils (the ones who had been written off as never going to amount to anything) in a council office. I experienced the impossibility of attempting to survive on Jobseeker's Allowance. I learnt that in cold weather turning off a stand-alone heater will only prolong the inevitable: the electricity still runs out. That hot water from an immersion tank is a luxury I cannot afford if I want a freezer and a fridge. I know what it means to be unable to afford a television licence or a television and still receive

endless warnings of imminent visits and court charges. I felt
the abject fear of not being able to afford car tax, an MOT
or insurance while still driving out of necessity. I know how
fast letters arrive for a single missed payment of council tax.
How it feels to pick up discarded cigarettes from pavements.
To get caught up in the bureaucracy of emergency loans
and food banks. I know the vulnerability of being desperate,
cheap labour, and the humanity of people who pay more than
expected. I learnt the subtle difference between the resolutely
altruistic and those who do not care. I experienced the joy of
being given free clothes, out-of-date food and gifts of neigh-
bourly generosity. On several occasions, when it got truly
hard, I learnt that sugar sandwiches help to stave off hunger,
as does thick, black instant coffee. I also discovered that a
kilo bag of flour lasts longer than a loaf of bread and a small
amount mixed with water makes a soft dough which, when
fried quickly over an open fire, produces a delicious, chewy
flat bread. That a 20kg bag of knobbly, bumpy, odd-shaped
carrots used for horse feed costs £2. The same amount of
money buys a net of onions and a sack of dried chickpeas in
similar quantities. I learnt that, under the cover of darkness,
potatoes, swedes and turnips can be pulled from the fields
and that hens' eggs are plentiful for a handful of change along
country lanes. I also learnt that, if stored correctly, and if I ate
only one meal a day, these ingredients will last a long time.
More importantly, I came to know that anything extra could,

with observation, ingenuity, self-reliance and a bit of hard work, be found for free.

Almost from the first moment I moved into the cottage I began to harvest wild food seriously. The knowledge gleaned from a lifetime of learning turned a walk in the woods from a delightful waste of time into a focus of the senses, and with this as with art, came a profound respite from the unresolved turmoil flicking back and forth in my mind.

The tracks and marks of folded grass where rabbits sat in the early-morning sun indicated more than the presence of a common animal. It told of frequency, numbers and their size. Without a gun or nets, trapping a rabbit is difficult. Rabbits can detect the small traces of sweat and particles of skin on the wire and will avoid the trap. To disguise the smell, I buried the wire snares in the earth for a week. It is easy to butcher a rabbit. The skin peels from the carcass like a leather glove from a hand; the two front legs, held only by a loose slither of mucus-covered muscle, shear away from the ribs. The tenderloin muscles along the spine are trickier to deal with. The muscular back legs are the choicest part. In a stew, rabbit flesh falls from the bone like pulled pork. A whole rabbit stuffed with two onions and a handful of wild garlic, thyme or rosemary and wrapped in foil cooks quickly over a grill in a log burner. I gathered the wood myself and learnt that each species has particular qualities, a particular character. Elderberry and willow burn fast and hot; other woods burn slow and cool.

When you're hungry, it takes only one charred rabbit to work out which wood to use next time. It was not only rabbits that I caught. Experimenting, I placed the traps along the trunks of trees and on branches. Grey squirrel has a similar taste and texture to rabbit, but with less meat and more thin shards of bone.

I followed the hoofprints and droppings of deer to quiet spaces away from humans. These places were full of natural life and abundant food. A heron in the same place at the same time each day has a reason to be there. Under the curve and whorl of water, shifting shadows revealed multitudes of fish. To catch them, I used flour paste, elderberries, worms, maggots, slugs and live insects. Becoming overconfident and for fun, I glued tufts of dog fur and feathers to a bare hook to imitate dry flies. Small fish with a light batter in a hot pan are ready in four to five minutes. Eaten whole, like sardines or whitebait, they taste of fresh ozone and soil. I stockpiled the excess. Frozen in groups of three or four, they last a long time but sadly lose the taste of nature when defrosted. I also caught bigger fish. A large pike produces several thick fillets of firm white flesh. Grilled, steamed or pan-fried, pike is a match for cod and easily surpasses the taste of Texas bass.

On the cusp of a full moon, fishing on the edge of an old weir pool, I hooked a huge sea trout. Monstrously strong, it took ten minutes to subdue and pull up on to the soft, grassy bank. Held to the moon, the fish glowed in a bejewelled, blue-silver sheet of organic chainmail. Fresh from the ocean, sea lice twisted and wiggled across its smooth, white stomach. Protected by laws designed to stop poor people poaching, only those rich enough to fish in exclusive private waters are allowed to take them. It took me two days to eat it.

Restless and unable to sleep, I would watch the early-morning escapades of otter and mink as they snaked up a grass bank to a nearby reservoir. Following them, I found the discarded shells and empty carapaces of crayfish. A dozen crayfish flash-fried in a wok and allowed to simmer in water until soft-shelled are filling and fun to pick apart.

An explosion of wood pigeons from conifers as a fox prowled below the trees pointed to nests containing eggs or young squabs. Pigeon is criminally undervalued, the flesh of a higher quality than fillet steak or venison, and even more so if cut from fledging youngsters. I also caught and ate other feathered creatures. Crows are surprisingly small when skinned, as are coot and moorhen. Ducks are almost impossible to catch without a hawk or a gun. Roadkill pheasant is far more productive than rabbit in terms of the meat it yields but can be flavourless in comparison.

Other than meat or fish, I found other free food to pick and take. The rings on grass, dark circles and marks, told of mushrooms before the fruits emerge. Ramsons, or wild garlic, has a short season and is common, as are rosemary, sage, horseradish and hawthorn. All provide light, fresh textures and flavourful additions to almost any meat, vegetable or egg. Others made a more substantial meal. Bypassed and forgotten, the Judas's Ear (a vile name that needs changing), a brown, slippery mushroom that grows all year round, can be cooked and eaten with flat bread. Twisting all over my garden and in the woods were wild hops. Eaten raw or boiled it was like having access to my own private field of asparagus. Ground elder, gathered in quantity and boiled, tastes like spinach. Burdock roots roasted on a fire are a match for any potato. Twenty feet from my front door comfrey grows in a wide green tidal wave. Shredded, they turn omelettes into bulky frittatas. The choicest leaves of comfrey have microfibres, a delicate fur perfect for holding batter. Fried one after another, they make a full meal with the tang of cucumber. Hogweed can be grilled or steamed. Wild daffodil bulbs look like spring onions and, added to eggs, they taste odd but delicious. Daffodils are poisonous. Within twenty minutes of eating them, I felt dizzy and vomited violently.

Never much of a gardener, I bought seeds and planted them tenderly. In genuine rapture and wonder, a tiny event, the perfect point in time, the runner beans, butternut squash, courgettes, tomatoes and peas grew thick and fast in the rich soil two feet from my door. I ended up with more than I could fit in the freezer. I got tired of eating apples, greengages, plums, damsons and blackberries. Freezing the fruit is easy, but defrosting them is not so good, as they collapse into a soft purée. The mush can be heated into a thick jam-like sauce and is more than edible in winter.

It is difficult to pinpoint the exact moment when I began to feel a sense of release. The change was gradual, organic, arriving slowly in layered periods of unrestraint as time rolled forward. I found it in the humour and peculiarity of life in all its forms. It was in the small victories when collecting food, in the slight moments of happiness walking in the woods and along the streams, it was in the strange beauty of the private scenes not normally observed in nature. It was the simple vivacity of life around the cottage.

Collecting wood, I watched a squirrel running from branch to branch, leaping across a gap in the canopy, missing and falling fifty feet into the edge of a pond with a thwacking bellyflop. Bedraggled and staggering up the bank, she crossed my path and ran straight back up the tree to try again. And missed again. A hibernating snake, accidentally disturbed, shitted on my hand in self-defence, a rancid, high, sticky stink impossible to scrub off. Other times I watched bats, with evolved sonar and supposed accurate echo location, collide into one another at dusk, or swing up into the branches three inches above my head to consume fat, furry moths. They were so close to my ear I could hear the crunch. I found this release in the spinning of my dog as she tried to pull grass from her bottom. In the spider in my window, sensing vibration on its web, shooting out and being beaten by a wasp. In the tiny pink slug touching and testing, delicately navigating the edge of a bottle for twenty minutes. On the lip, just crossing to fresh leaves and food, a bird flew down, smashed it on a flowerpot and ate it. I felt it in a fledging woodpecker on a maiden voyage, a stuttering, half-feathered flight, landing straight into a stream. I lifted the youngster from the water and she attacked me, beak open, hissing, vomiting water and beetle shells. It was in the low-pitched cheeping of kingfisher chicks ensconced deep in the bank of a stream as a predatory mink swam in circles below the entrance. The laser-line-blue parents skimming and screaming across its back. Or in the scattering of baby dormice fallen from a nest, naked in newborn pink and the size of fingernails. In full view, I watched their mother picking each one up and carrying them back to safety. In the five ladybirds side by side in various stages of change, from the soft-spiked, pillow-shaped youngster to the hard, shiny-shelled adult. In spotting (and quickly walking away from) a pair of semi-naked sixty-year-old humans reliving the summer of love and bare-naked frolicking in a secluded corner of a meadow. It was in clearing brush and rotten logs from around the cottage. From the fug of compost and slime, small, thread-like red worms and grubs fell to the floor. A flash of brown from the right, and a small robin darted in, tilted her head and skipped across the earth, picking up another free meal. She followed me around all day, then the next, and the day after that. Growing in confidence, she began landing on the handle of the shovel, then on my boot. As the sun set and I was weak from digging, the shovel merely scraping soil and bouncing off rocks, she landed under the blade and remained. I reached down and folded my hand around her warm body, gently picking her up. In her beak a small worm twisted, trying to free itself. I placed the bird on a branch above my head. She looked at me and gulped down the worm.

Around mid-June in the second summer, restless with bad dreams and old memories, I padded downstairs, rolled a cigarette, poured a cup of cold tea and sat outside. The dog bumped about, annoyingly nudging a wet nose under my arm, as the sun rose behind an elongated black smudge of the far treeline. Most summer dawns are stunning. This was no exception: it was a cliché unfolding in vivid, delicious detail.

From the garden I waded out into the crops wearing only boxer shorts. The corn, three quarters grown and young, still had a white, waxy surface. The dog was a shark swimming in a sea of green. Delicate cobwebs swathed the tops of the corn, covering fifty acres with a transparent gossamer sheet of undulating, pulsating bubbles. A million spiders spinning webs in the night, a million spiders climbing up to dry their washing in the dawn. The morning sun caught droplets of dew on each web and sections of the fields burst into patches of paparazzi camera flashes.

Sweeping my hands across the soft, wet webs, I remember having absolutely no idea what day of the week it was. It could have been a Monday, a Wednesday or a Saturday. The days had leached elastically into one long moment. It did not matter, anyway. I had nowhere to go, no deadlines to meet. I had food, fire, a field of flashing spider webs and two paintings to finish.

In those few minutes of freedom, between not knowing, knowing and not caring, an old, familiar and euphoric feeling hit me. It was the same sensation I had felt numerous times when travelling. I had felt like this when wandering out into the dawn in Pakistan and sipping tea, watching the boy and his father arrive with the trapped musket. As I stood in the corn and the dawn drew larger over the horizon, I knew my day ahead was a concentrated and amplified version of the splendid isolation I had felt in the hut in Slovakia and when drifting down the highways of Texas. It was the freedom I had felt watching the tip and tuck of descending falcons over the plains of South Dakota and when spiralling through the bubbles with the water snakes of Croatia.

It had taken me nearly two years to rebalance. I should have stripped back to the bare essentials, emancipated and simplified years ago. Instead, by misdirecting, misunderstanding and denying my own instincts, I had failed spectacularly. I had lived a life of self-created discord and made terrible choices, trapping myself in the rules and worlds of others, hurting them and falsely believing I could fulfil their expectations. The truth was, I was incapable of doing so.

Here, in a broken-down cottage with no recognizable materialistic possessions and alone save for an undiluted relationship with nature, I was existing without crippling fear or anxiety. I was laughing at woodlice under logs and weevils in the flour, happily weaving the natural world deeper into the fabric of my day-to-day existence. Instead of guilt and shame, I felt an inestimable pride and sense of purpose and, although the living was hard, it was also playful, child-like and fun. I had regained an independence of spirit, I felt a creative cause and effect between my actions and my life. I was controlling the parameters in a constructive way and my unusual behavioural characteristics were a strength. In the right context, the quirks and tics of my biology, my frailties and faults, were of worth, allowing me to exist in my own unique and peculiar way.

I was free to be myself.

I felt as free as a bird.

Freedom and flight are inextricably linked. Flight is a momentary escape from gravity. Flight has levity, moves in any direction, free to travel, in migration, nomadic or on passage. Flight is free spirit, whirling, cavorting, hunting and flying for fun. As a falconer, I knew this as a concrete experience, not as a symbol or a metaphor. I lived alongside hawks and travelled great distances to feel the freedom of flight in equity. Throughout my self-imposed exile and recuperation I knew that sublimating the innate freedom of a hawk as a cure for my pain, to steal it as a distraction from the complexities I had created, would limit, dilute and fundamentally change my relationship with birds of prey. Any semblance of unbalance, lack of time, sadness or pain on my part would mean functioning at half-capacity. I knew a hawk to be simply too important to carry and cure human folly. A hawk deserves, requires every ounce of clear-minded connection and energy. Training and hunting with a hawk is demanding, focused, a labour of love. A love not found in romantic abstraction, in words or thought, but in deeds. It is a love expressed through vocation and action, in the details of care, in a daily routine, in observation and ownership. For these reasons, I had not owned a hawk for just under two years and, although I had sent him drawings, cartoons, presents, cards and letters, I had not seen my son for a similar period of time. As a falconer and as a father, I had been consistent in only one thing: my absence.

With this increasingly stable sense of self, it was only a matter of time before the situation turned. In my second summer I received two phone calls from different falconers. Both needed help with injured sparrowhawks, a female and male. A complete set, both wild, both raised by nature. One required help immediately; the other would be ready at a later date.

Now old enough to talk, to understand absence, old enough to ask questions, my son also wished to see me.

I agreed to both.

4

The Ascent

I wake early, get up and pad down the stairs. Swinging the door open, I walk to the kitchen, pour a mug of tea and sit on the front step. The shrill, cross-cut sounds of the dawn roll into one another as the sky begins to lighten. I hear a muffled thumping from upstairs: Etta, the dog, is fighting to extricate herself from under the duvet. She staggers down, bowls past me and bumps my shoulder, chasing invisible scents up the path.

The first injured sparrowhawk arrives today and the prospect precipitates a running bundle of nerves, a deep internal tension. I think ahead, my mind a looping list of equipment needed to keep her safe: food, gloves and drugs, perches, leather, lure, and leash, bells, batteries, scalpels and swivels. Some of these things I already have; others I will need to order.

Moving back through the cottage, I stop and flip through old magazines, look at photographs, replay short snippets of film and skim falconry books. Surrounding me are the ephemera of nearly fifteen years of hawks and travel: removed by vets from injured birds, small tubes of infections, waxy substrates, splinters of blackthorn and snapped feathers make a macabre museum of accidents and injury. The handmade bells of Mohsin Ali are tied tightly together with string and are suspended from a hook in the kitchen. The brightly beaded ceremonial falconry leash presented by Salman, Haider and Ghulam is coiled in a loop on the arm of a chair. The sofa is covered in the geometric-patterned blanket designed, sewn and given to me by the tribal women. Falcon hoods, coated in cobwebs, hang on pegs above the fireplace. The dried skins and feathers of rabbit, pheasant and partridge are loosely nailed above the doors. A shed golden-eagle feather eighteen inches long is tucked into a hole in the wall. Hollow eggs are mounted on card or kept cool in the fridge. Displayed on the walls, or stapled on oak beams, are drawings, paper clippings, paintings and photographs of hawks, eagles and falcons in flight. Stored in no particular order, books covering subjects as diverse as incubation, fishing knots, canning peaches, wild herbal remedies and dry-smoking meat are piled high in columns or stacked on shelves.

Alongside this falconry-themed detritus, pinned up, scattered about, some framed, others on curled pages of A4 paper, are multiple drawings by my son. His art is unrelenting, a creative cacophony of weird sketches. Fat dogs farting, aliens and frogs. His mother with a hat or a frying pan on her head. Several smiling figures with elephantine legs and no feet have thin arms and circles for palms and hold their hands up in surrender, or perhaps in a friendly wave. A giant oblong jelly-fish is roaring with laughter, a huge grin with teeth the size of tombstones.

Spidery writing, difficult to decipher, a title or an explanation, walks its way off the edge of each page. My favourite picture by far is an amalgam of paint mashed and smeared across card – a psychedelic hawk with wobbly plastic eyes. The work exists at a level of freedom all happy children possess.

Etta crashes back through the door, skids across the floor, distracting and snapping short the tangled thoughts of a new hawk and the memories of my son. I finish my tea and get dressed.

After a short drive across Shropshire, at precisely 9.32 a.m. I set eyes on the first rescue sparrowhawk. She is the size of a kitten and is sitting in a plastic tub, lazily collapsed and spread across newspaper. Overconfident in her abilities, she clambered from her nest and out across the branches before tumbling to the ground. Found by a well-meaning dog-walker, she was passed to a wildlife rescue centre, who in turn gave her to the falconer I am talking to.

She looks as if she has been stitched and sewn together with oddments of different animals, built in a hurry by a worker from Jim Henson's Muppet studio. She is both beautifully ugly and interestingly repellent. A lizard-like Gonzo reptilian mixture of legs, wings, scales and wonky feathers.

Waking to the sound of a new voice, she stands, yawns and stares directly into my eyes. Almost an adult, close to being fully fledged, she retains the awkward, uncoordinated clumsiness of a young hawk. Twisting her head round, she nibbles at new feathers, setting free white fluff that floats about like dandelion clocks puffed off a stem. I open the boot and reach down to pick her up. She is fairly heavy, maybe nine or ten ounces, and very warm.

She begins to struggle, flailing feet and talons grasping and clasping at air. I feel the poky bones beneath her skin and the bristle-stiff brushing texture of feathers across my palms. Her chest expands and contracts rapidly. I listen to her breathing. The air from her lungs escapes clean and clear into my ear: the noise is healthy, arriving in a regular, soft, meat-scented hiss. Her left foot suddenly springs out and latches to the lip of my cap, pulling it from my head. I place her back into the tub and gently unravel her first 'kill'.

Thanking the falconer who has delivered her to me, I carry her to the passenger side of my car, open the door and place her tub on the seat.

On the drive home she rustles about in her box, is quizzical but relaxed. I begin to run through a list of names in my head.

Wishing to keep it simple and knowing she will be released,
I settle on the obvious, the functional.

I call her Girl.

*In the same month as Girl's arrival I stand nervously in the grassy
car park of a National Trust property. A grey/green jeep turns
through the gate and approaches. A small boy I do not recognize
bounces up and down, trying to escape the confines of the car.
A greasy palm-print smear appears across the windscreen. Before
his mother can stop, he launches himself through the open door,
runs and jumps up with full force, grabbing me tight. I hold him
tighter, squeeze him until he lets out a grunt and a giggle. For such
a lithe human, my son feels heavy, hot and extremely bony. I can
hear him breathing, feel his heartbeat. He smells savoury sweet,
a mixture of Mini Cheddars and jam sandwiches. I am surprised
at his emotional warmth. I had no idea I would elicit such a
strong reaction.*

*We walk around the park for a few hours before finally sitting
down on a soft blanket and eating some food, a picnic of sorts.
I feel a strange, disconcerting recognition, a feeling absent when he
was smaller, absent when I last saw him. It has been a long time
coming, and I start to relax. As I look at him, I see the vaguely
traced shadows of my younger self. He is different in shape and size
from me but, as he turns and moves, I recognize our similarities
and our differences.*

BEN CRANE

His skull is large and his ears stick out. His eyes are dark brown, like his mother's, whereas mine are electric blue. Like mine, they are wide and intense, almost almond-shaped. His face is thin and he has a shallow, pointed jaw. When I make him laugh I can see that his front teeth, like mine, are slightly angular and twisted. From the canines back they are perfectly shaped, similar to his mother's. When he agrees with something I say, he nods and smiles like me. His vocabulary is broad and surprising. He is chatty and the intonation of his words is an almost perfect echo of my own. He looks and sounds like an attractive elf, and this recognition, this resonance of myself, makes him strangely beautiful to me in a way I did not feel when he was a baby.

We begin playing with his toys. Without thinking, my son calls me Daddy. He was too young to use this word before. This is the first time I have ever been called Daddy. At first, I do not respond. I do not consider myself to be a father; the word does not correlate to who I think I am. I try to shut these thoughts away, and we continue to play normally. But it jars, 'Daddy' feels weird. I still have no idea what a father should be. I am out of my depth and somewhat scared. My son seems to know more about who and what I am than I do. That this small boy attaches himself to me so easily, anoints me and unquestioningly confers on me the role of a father is an act of faith and survival I find startling. The power of his attachment is alarming.

After a few hours it is time for me to leave. As first meetings go, it was perfect. I feel good. Then halfway home an unexpected sadness washes over me and I experience a sudden exhaustion.

By any level of objective reasoning, I am a rubbish dad. I make no sense to myself. Meticulous and naturally confident in my relationship towards an entirely different species, I possess a restless confusion and sense of failure towards something I created.

The whole situation feels totally unnatural. It's like looking at two ends of a piece of string and the messy middle is balled up at my feet and fails to connect. I begin to miss him profoundly. I begin to miss him in a way I have never missed anyone before.

At the cottage I place Girl near the fire and leave her to unfazed snoozing. While she sleeps, I walk to the freezer and remove her dinner of two small yellow cockerel chicks. I let them defrost in a plastic tub and pick up my mobile phone.

The legal possession and ownership of a wild hawk in England is a convoluted business. It is not sufficient to simply take possession of Girl, nurse her back to health and then set her free. She is heavily protected by law. Rehabilitation carries legal responsibilities and complexities which, if sidestepped, ignored or not followed, could lead to prosecution and a custodial sentence. I dial the number for the department of Natural England and explain that Girl is in my possession, where she has come from and my intention to release her. I phone the local wildlife police liaison officer and invite him to inspect my premises. By late afternoon Girl is correctly monitored and declared, her legal position vis-à-vis various government departments secure.

Moving back to the kitchen, I peel the fur off the pliably soft chicks and carry Girl's first meal to her. She immediately stands up and flaps forward, grabbing the chicks in her feet. She pulls and tears at each chick. The first few mouthfuls are swallowed in an aggressive manner, a kneejerk reaction. I know she is not hungry, but her instinct to feed overrides all else. Just as quickly, she grows bored, stops eating and, leaving three quarters of her meal, bounces back up on to her perch.

In a week or so I estimate she will be fully feathered and able to fly. It would be easier and less time-consuming to simply release her then. The direct contact with humans has most definitely saved her life, but it has also stalled her natural upbringing. If I let her go so soon, she would be vulnerable, young and immature. If Girl is to survive, before I can release her I will have to train her, make sure she is in good health and can hunt independently of humans. Unfortunately, a significant proportion of Girl's natural quarry is off limits. Like Girl, they are protected by law. Running concurrently with all the other legal strictures of rehabilitation, I return to the phone and am shunted between different government departments as I request and apply for a licence to temporarily hunt a small number of birds as part of Girl's rehabilitation.

There are further complications. Being wild-bred, Girl does not have a legal breeder's ring or any legal documentation.

She will be impossible to insure. Until the time of her release, all medicine, food, housing and other sundry costs will have to come from my own pocket. If she falls ill or is injured, her treatment could potentially cost thousands of pounds in specialist vet fees. If she breaks her wing, blinds herself or becomes disabled and cannot be released, her protected status makes it illegal to euthanize her. I will therefore have to feed and house her for the remainder of her natural life. Girl could live for up to ten years in captivity, and the cost of her upkeep would continue to mount as the years passed. Over and above the trivialities of finance, my deepest fear would be keeping Girl in a depressing half-existence, trapped, caged and contained in an aviary for the rest of her life.

I return to Girl and now she has eaten the remainder of her dinner, she emits a satisfied, low twittering and stares at me with piercing eyes of green-grey blue. Shuffling her feet, she lifts her tail, vibrates and explodes a slice of faeces (a mute) three feet across the room. With unerring accuracy, it loops cleanly up over the muzzle of Etta. Girl wobbles back around, commences to flap and lift a few inches above her perch, lands abruptly, then makes a snickering sneeze. She is utterly unconcerned about her future.

During the first half of living life with Girl a relaxed sense of fun permeates the cottage. Riding high on the fat of the land, nothing is required other than eating, flapping and sleeping. On a full diet she quickly morphs from the fragmented, odd-angled juvenile into the smooth outlines of an adult hawk. In under two weeks she is now the size and shape of a delicate, refined blown-glass vase, all traces of white and grey fluff gone, exchanged for an overall light, tree-brown coloration. In scale-on-scale layering, each thumb-sized feather on the arc of her shoulders has thin, inverted C-shaped bronze edging. The underside of her wings and tail are pale white and striped with bars of grey. Her chest pulses with tics and swirls of autumnal coffee over pale cream.

When touched, Girl's flight muscles feel rotund, the bones of her legs and feet toughened to thin rods of steel. Her eyesight has developed exponentially. She locks on to the slightest movement in the cottage. Moths, wasps, flies, scuttling beetles, the mop, the flames of the log burner and, particularly, the dog, are followed with avid interest and increasing determination.

One morning I find she has caught a mouse. Her neck is bulging, there is a small patch of blood and a tuft of fur under her perch. For the rest of the day she makes contented beeps and thin tweets of self-satisfaction. The death of the mouse is significant and points in one direction only. Her drive to kill means it is time to start her training.

If truth be told, I have been putting it off for a few days, and my growing sense of unease begins to unfold like a flower.

The build-up to training any hawk is always tense, like a cord tugging and attached to the solar plexus, pulling me towards the inevitable. When I force change, when training begins, I know Girl will reject and vociferously protest against the process. Initially, she will not waver and one mistake on my part may mean disaster. The fluidity of these first few weeks is never without issue or indeed brief moments of grave despair.

When training my first sparrowhawk, on her very first free flight she simply flew off into the woods and became a small dot, dissolving invisibly through light and shade. My world condensed and collapsed with a sense of utter helplessness. Any romantic thoughts of freedom and flight were quickly squashed and swapped for a violent urge to climb up into the sky and punch the sun. I entered the wood, looked about for a while and then did the only thing a falconer can do when a hawk is lost. I started whistling at trees. After what felt like a lifetime she moved and, following the sound of her bell, I chased her around for several hours before she returned. It was close. She could have been lost for ever.

The first falcon I trained, a little merlin, followed almost the same pattern. After a week's preparation I removed his training line, set him on a fence post and called him in. He flew from the post, across my shoulder, and curled straight up into the sky. I watched helplessly as he slid through the air with the rolling ease of a consummate professional. His entire physicality changed, as if to say, *So this is what I am for*, and he began flying with surging, unexpected energy, reaching hundreds of feet in three fast-flying turns. He was so far up and so far away I lost sight of him. I was devastated and meekly followed in his general direction, swinging the lure and calling. Five minutes later he reappeared high above me, dive-bombing swifts and swallows before landing on the roof of an old church. Climbing the bell tower, I skidded about on an 800-year-old roof roughly 150 feet above the ground and grabbed his legs, pulling him back to earth.

The recurring problem with young birds of prey is that they lie. They pretend to be ready for free flight and they are convincing. The temporary loss of a hawk is therefore almost a certainty until a routine and trust have been set in place. This takes time and delicate observation.

My son lives a long way from me. Three motorways and a five-hour drive. He has his own routines and is a homebody. The journey he took to meet me with his mother involved a massive traffic jam on the M25. He tells me with humour, good grace, but in no uncertain terms: 'I am not doing that again'.

This makes me laugh. I know exactly what he means.

On my first extended visit I stay in a motel near his house, spending time with him during the day. Towards the end of the first day I start clock-watching, time bends, speeds up. It is upsetting coming and going; it is neither one thing nor the other. I feel awkward, under pressure to perform, to cram everything in, and the time to leave arrives far too quickly.

On the second day my son asks if I will stay the night. I tense up, absolve myself from committing to anything: 'It is not up to me, mate.'

His mother says she does not mind.

I wish I had said no.

It is easy for adults to pretend to be nice to one another in front of a child, but it's far harder to maintain over time and in close proximity. I know from my own background that the history of fractured families is full of venom. Emotional poison flows through intra-family relationships when they split. There are always rolling recriminations and the dragging up of misdemeanours: he said, she said, you did this, I did that. That's just how it goes. It is what I am expecting.

The whole day I am uncomfortable, my antenna high, my guard up. I am waiting for a comment, a look, a raised eyebrow. I am waiting for any sign of righteous animosity, anger or negativity. I listen to what my son says, waiting to hear him accidentally repeat some snide comment fed to him about how feckless his father is.

I hear nothing of the sort.

I see nothing of the sort.

I am perplexed.

I begin to see that, without the pressure and expectations of a fixed relationship, our dynamics have changed for the better. His mother is wholly different, less controlling, less stubborn, less critical. The grinding tedium that accompanies the end of a relationship and the hatred that so easily festers are non-existent. There are no bitter broken words when I put a fork in the wrong drawer, there is no splenetic rage when I eat the last biscuit, no cold annoyance at my very existence. That I breathe is not an issue.

I begin to think that maybe she sees me in her son. Through him, maybe she has a better understanding of who I am. It is more likely that I do not figure in this change of attitude at all. The love for her son is clear. It overwhelms all else. Anything that may hurt him – any negativity, words or actions of the past – have been banished. All that matters is the here and now. Then again, I too have changed. I am far more relaxed, less argumentative, less angry, far less fearful and a lot less destructive.

I watch them intently. Listen to them interact. She has done an exceptional job building their lives. She is not a victim of my weaknesses or my behaviour. She is not a pushover. She does not suffer fools gladly. Her job in London involves managing multi-million-pound budgets, organizing a workforce, placing people to perform to the best of their abilities. She has applied the same determination towards her son and is protective and generous. The relationship she has created with her son functions perfectly well, with or without me. Nothing I do will interfere with what she has created. She has her own independence and freedom, and my son thrives because of it. I can see clearly how the structure of their lives functions. I can see where I fit in, what my purpose is. I feel valued. I am able to participate.

When he is in bed she tells me that what happened doesn't matter. What matters is how we move forward. It is easier to communicate with her now. We say the same things using different words. We agree not to argue. It is far better to be kind than to carry the exhausting weight of hatred for the next forty years. It is that simple. That black and white.

In the morning my son wanders downstairs and climbs on my sofa bed in the front room. His hair is standing on end, he looks a bit crumpled, a bit bleary-eyed. He nestles under the duvet and we watch cartoons. When his mother wakes up we have a cooked breakfast sitting in the sunshine.

All I want is space and time to learn the one thing that does not come naturally to me. How to be a father. This is what I am given.

On the first day of Girl's training I place her on the top of a fence post, walk a few yards from her then call her to the glove. She lands hard and snaps off her back talon. Thick, deep-red blood flows freely, smears across the leather thumb of my glove and drips to the floor. This is very bad.

I run towards home with Girl on the glove, blood oozing up through my fingers as I frantically try to stem the flow. Restless and upset, she flaps about, swinging heavily around on the end of her leash. When she is reclaimed, replaced and stable, I notice that a second outer-right talon is split and frayed at the end.

Once inside the cottage, the keratin coating on each damaged talon slides away, revealing a small nub of pinkish-red raw nerve endings on each toe. The pain must be immense – like tearing a nail from a thumb and forefinger and being expected to complete a Rubik's cube. I search frantically for the first-aid kit as blood continues to fall from her wounds, dropping with a *tip-tap* to the floor.

Iodine is a remarkable compound, a potent antiseptic and caustic, perfect for chemically cauterizing wounds. Shaped in spiky hexagonal crystals and purple in tone, tipped from a tube in panic it bunches up on a table like iron filings on a magnet. In crystal form it is highly concentrated and, if blown across the room by the draught of a hawk's wings, stains wooden floors a deep puce. If an errant crystal is mistakenly eaten, the taste is stark and wincingly sour. Poured into a wound, iodine stings like lemon juice in a paper cut.

I dip a wet cotton bud into a small pile of iodine and press it with force on to the nub of Girl's exposed toes. I tense up, expecting her reaction to be one of violence. Instead of explosive anger, she lowers her head and gently nibbles her wounds, as if scratching an itch rather than reacting to a burn.

Inadvertently, she has taken a tiny fleck of crystal in her mouth. Her arrow-shaped tongue turns deep purple. I watch as the taste hits the back of her throat. She pokes out the angular tip in disgust but remains calm. It takes a few minutes for the blood to stop leaking out from under the iodine. When it does, I place Girl in a darkened room to keep her calm and quietly close the door.

Like all hawks, Girl's blood is highly oxygenated, so it has a viscous consistency similar to watered-down strawberry jam. It takes a few minutes to wash my hands and wipe away the trail of blood through the cottage. I find spots and splatters on the front door, the step, the path, the garden gate and half-way down the track back towards the training ground. It is the heaviest blood loss I have seen in any hawk.

I check on Girl periodically through the night. By dawn, the iodine crystals have dried to a black crust. I peel it off, check for pin-prick bleeding and apply a second layer. Once this second shell falls away, I spread a thin layer of liquid skin over the ends of her toes to act as a flexible, porous barrier against further infection.

I spend several days impatiently waiting, internally arguing about the merits of flying a hawk with two missing talons. All arguments cease when Girl stops eating and starts flicking food across the floor. The spiralling disaster of ill health continues to turn. Due to shock or simple bad luck, the soft, warm edges of her mouth become a breeding ground for a yeast infection called frounce.

Running down the blue/grey and pink edges of her throat, small white speckles blossom to the size of grains of rice. These soft pillows of fungus act like a fish bone trapped in the throat. They block her windpipe, inhibiting her ability to swallow.

I catch a sour smell when she breathes, the rotten tang of old flesh or metallic, stinky fish. The infection is slowly turning her breath rancid.

Without medication, the frounce will continue to multiply and Girl will starve. I roll her up in a cloth and gently tug at the feathers below her lower mandible, opening her beak. Using a moist cotton bud, I scrape the effluent from inside her mouth so she can swallow. I mince her food to a sloppy pulp, add a drop of water, avian vitamins and a course of crushed-up, inexpensive anti-protozoal drugs from the vet. It takes a few days of throat-scraping and drug-laced soup before the frounce begins to recede.

By the end of the second week Girl does her best to eat four whole chicks a day, but they sit awkwardly, slip through the gaps left by her missing talons. She also finds it difficult to stay on her perch properly and keeps sliding around on it. She is hysterically restless. To stop her bursting open the scabs, the only option left is to raise her weight to its highest, healthiest levels and set her free in an aviary. If her talons fail to grow back or are misshapen, or the base of her toe damaged to the point that it cannot regenerate new talons, my deepest fear of Girl living her life in captivity has been realized.

I am heartbroken for her.

I am heartbroken for myself.

At the far end of my garden, past the pampas grass, almost into the wood, is a small area about an eighth of an acre in size. It's overgrown, half in sunlight, half in shade, and in it voles shriek and fight, snakes sunbathe and a tawny owl nests above a hive of wild honeybees. It is a private place, secluded, ankle deep in fallen acorns, chest high in nettles and surrounded by plum and damson trees. I strim back the nettles, clearing a space, and build Girl a large, secluded aviary.

Hawks do not like being caged. If the design of Girl's aviary is wrong she will haphazardly crash about, relentlessly flying into the walls and corners. A badly designed aviary will cause her to smash feathers, split her beak or, worse still, break her bones. Before I put the roof on Girl's new home, I stretch several layers of soft mesh netting across the top and staple it in place. If startled, Girl will gently bounce off the netting without injury. Along the right side, facing out over fields, I screw vertical rows of white overflow piping across large squares cut away from the walls. This allows free-flowing air and sunlight to stream between the bars. In any attempt to escape, Girl will simply grab the piping and slip safely to the floor. For a fresh supply of flowing water, I drill a hole in the bottom-right-hand corner and push a hosepipe down into a large circular bath. Several thick perches are attached to the walls and across the centre of the aviary. A small ledge and feeding hatch complete her new home.

I release Girl into the aviary, duck down and hide quietly

behind a damson tree. After initial suspicion, she dances about on the different perches, flies to the floor and has a bath. Wet and spiky, she hops back up near the window and calmly preens and cleans her feathers. Sneaking around the side, I drop diced pigeon breast coated with calcium and vitamin powder and what remains of her medicine through the feeding hatch. She is safe, settled and comfortable in her home. I have done as much as I can. Other than feeding her daily, the rest is up to her. There is nothing more I can do to speed up her recovery: nature cannot be rushed, and a talon takes eight months to grow back. I will have to keep Girl contained through the autumn and winter. It will not be until next spring that I find out if her rehabilitation can continue and whether she can be released.

My son and I are in a church in December, and it is freezing. Cold, and echoing in a way that only very old churches do. My son is semi-naked and excited. He twists and turns, laughing in my lap. Trying to get Nebuchadnezzar into his costume, preparing him for the birth of the baby Jesus, is proving more complex than it should. A small sheep and a camel wander across the stone floor-ing and ask if I am Nebuchadnezzar's dad. Happy with the answer, they wander back to their respective herds.

    This is the first time I have been to his school, the first time I have participated in his education. As a teacher, I visited a lot of primary schools, took part in hundreds of workshops, plays and productions and met dozens of teachers. I knew primary schools inside out, was confident and in control. As a father, my role has changed significantly.

When I first arrive at his school I stand with my son's mother and watch the teachers rushing about, collecting fluorescent jackets for the pupils to wear on the walk to the church. I feel the stigma of being an absent father acutely. I know the conversations, classifications and how the reports are filled in. I know the social dishonour of absence and carry my self-created wound close to the surface. I am in a raw, nervous state as the parents, teachers and children group up and crowd together. I am deeply self-conscious about being here, have a fight-or-flight level of fear. I want to escape. I look for release, a way out.

My son rushes up, his energy breaking my cycle of thoughts and feelings. On the walk to the church, adults smile inquisitively but remain silent. The children have no such reservations. This is the first time I have met his friends, the first time I have seen him interact with other children. 'Who are you?' 'Is he your dad?' 'I didn't know you had a dad.' Their fearless, beautifully honest questioning continues. Holding my son's hand, I listen to him explain with pride who I am. He lacks all anxiety, speaks and explains fluidly. I remain quiet and let him talk. He gives me my escape route, and I thank him in my mind.

In the gap between dressing him and waiting for the nativity play to start, his grandparents arrive. I have not seen them for a long time and I feel deep embarrassment. Other parents take out their phones and start filming. The church is filling up. My mind begins spinning again. I feel the pressure of people close by, see bright colours and hear loud conversations.

Twitching, nervous, I move right to the back, to a place of safety. The play starts and I am swept up. I forget about myself and escape into my son's world once more. He sings and sways and dances, is freely artistic. His urge for creativity, for performance and for fun is mesmerizing. The children deliver their lines, overlapping and crashing into each other's words. Hats slide from heads. My son is pinching and pushing Melchior. He catches my eye, and I pull a face. He misbehaves splendidly and is told off. In the half-hour it takes for Jesus to be born, I finally know what it feels like to live vicariously through my own child's enthusiasm and lust for life.

In falconry, dogs are sacred. A hawk, a dog and a human coexist vicariously through their own separate senses. They work together through amplified sight, amplified smell and the conscious coordination of the human mind. A working dog, like a hawk, is a gift beyond mere companionship. They earn their place in spectacular ways.

Etta is a pointer. Specifically, a Hungarian Vizsla. She has a russet-ginger and gold coat and a thin, sculpted face. First bred 2,000 years ago by the hunting Magyar tribes of

Hungary, Vizslas are said to be the oldest pointing dogs used specifically to aid falconry.

Etta is athletic and powerful, and her chest is deep; her lungs take up at least half of her body. Side on, she looks like the outline of a rasher of back bacon. She will run all day over tough ground searching for quarry, and her sense of smell is one of the most beautifully evolved adaptations in the whole of nature. When she detects a pheasant she turns, locks solid, stationary, and points at it with her nose. On command she will edge forward and flush it for the hawk. Dogs of Etta's calibre deserve to have offspring.

With winter fully upon us, Girl alive and recuperating in her aviary, and after sixty-five days of gestation, I help Etta give birth. Other than my son's mother, this is the only other living creature I have seen giving birth.

In the late evening on the due date, I close the curtains, stoke the fire and wait. Etta has been restless for several hours. She wanders upstairs then comes down again. She climbs into her whelping pen, tears at the soft layers with her paws and starts a low panting. She climbs out of the whelping pen. In an effort to get things moving, I take her for a walk up the lane and back. On her return, she hops back into the whelping pen, tears at the blankets once more, turns and sits down. I watch the sides of her body spasm and contract. I begin to panic. I realize how inexperienced I am. I am alone with her in the middle of the night. Streaking thoughts flow through my mind, horror stories of breech births, dead puppies and dead dogs.

A newborn puppy on the edge of the womb and world plops out like a soft-poached egg. When the first pup is born she remains motionless inside her cocoon, as if stillborn. I feel sick with disappointment. With a mounting sense of revulsion, I move to fetch a bucket to put the lifeless body in. As I do, Etta starts licking it vigorously and breaks the sack with delicate nibbles. The air hits the pup's lungs and it twists alive, bending back on soft bones like a caterpillar unfolding after falling from a leaf. It is incredible. The little pup is blind but not helpless. Peeled clean and grunting, the furry orange piglet does not hang about but wriggles over the soft bedding and latches on to Etta's nipples, supping life into its tiny stomach.

Each little Vizsla arrives in the world the colour of a packet of Plasticine rolled together: an amorphous grey blob with blue veins and swirls of flesh tones. Each pup is small enough to fit in the palm of a hand, is broad bean shaped, and smells of warm mushrooms and damp earth. When the moisture evaporates, it draws tight on my hands and fingers like dried gelatine, and the smell changes to the tang of raw liver: a light, watery metallic oxide. It takes eight hours for all six puppies to arrive, and they do so with minimal fuss.

When the last pup is born, cleaned up and feeding, it is close to dawn. Etta is exhausted. I hand-feed her small pieces of boiled chicken, give her some water and take her outside for a wee. When finished, she bolts back and scrapes at the door. She hops back into the whelping box, spins in a circle and flops down. A Grand National of pups squirm and roll across the blanket towards their mother. They are unforgiving and fight, climb and kick each other out of the way to get the best position. She gives them all a protective nudge, licks them clean and snoozes while they feed.

I am tired too. I feel empty, a hollowed-out void. All thoughts and fears for their safety are now blown away, replaced by the soft wash of endorphin love. It is one of the most moving things I have ever seen. On the sofa, I drift between sleep and dreaming, the tiny squeaking sound of suckling mixing with the click and crackle of the fire. The cottage seems to swell around us, soft, warm and protective. I fall through my mind, slowly dropping off to sleep. When I wake, I count the pups. Etta has given birth to a seventh all by herself. Lil Titch is half the size of the others, the runt of the litter.

Unable to keep up with his brothers and sisters, Lil Titch fails to feed and over two or three days becomes weak and floppy in my hands. I begin mixing formula and feeding the little orange grub by hand every few hours. My clothes and the chair take on a strange cheesy sour-milk smell which I find strangely comforting. It reminds me of my son. It is touch and go, but I save him. Lil Titch's ability to bounce back is joyful: after a week he is happily kicking his brothers and sisters out of the way to feed.

On the day Etta's pups go to their new homes, my new puppy arrives. He is a little black Cocker Spaniel with a white stripe on his throat. Unlike Etta, he does not point. His job is to crash through bramble thickets, nettles and thick cover to find the pheasants she cannot. Prompted by his patterns, I toy with the name of Flash. I think it would suit him. Bold and confident, he wanders in through the front door, walks straight up to Etta and begins suckling what remains of her milk. She looks at me. In heartfelt response I say: 'I guess it's not over yet, my love.'

I tell my son about the new Cocker Spaniel. My son was raised around dogs, in particular a giant Alsatian called Stevie. Since Stevie passed away my son has been pestering his mother for a Golden Retriever. He has a natural proclivity for animals but, like all children, his level of commitment and care wavers. So I bought him an electric dog, one that breathes, one that needs batteries, but does not need feeding or walking. Over time his Golden Retriever stopped breathing and my son failed to put new batteries in it. As a halfway house, a compromise, I tell him he can have the new Cocker Spaniel. It is his, and I'll bring him down every time I visit, but I'll look after him. I ask him to think of a name. My son comes up with two. Ironman and The Flash.

For obvious reasons and with unconscious synchronicity, we both decide on Flash.

When spring arrives, just prior to removing Girl from her aviary I am allowed to collect a second rescue sparrowhawk. Taken as a thumb-sized chick from the wild, his tiny legs and underdeveloped feet allowed a breeder's ring to be surreptitiously slipped over his foot. Whoever illegally stole him was attempting to give the impression he was registered. In fact, the printed numbers and letters around the aluminium band were for a kestrel, a deceptive ploy discovered only when he was found and handed to an experienced falconer and law-abiding breeder.

When I meet him the little musket is fully grown and easily the most attractive bird of prey I have ever seen. Exactly like the trapped musket in Pakistan he is a third smaller than Girl, he's less than a foot tall and his legs and toes are as thin as toothpicks. His eyes are crystal clear and the saturated colour of deep tangerine. As bright as a parrot's plumage, his neck and cheeks are a delicate mixture of burnt earth, red clay, flecked carrot and iron ore. His back and shoulders are a gradation of dark blue to slate grey. In contrast to Girl's pale white chest, his is the tone of double-thick clotted cream, with jagged, zigzagged, flashed bars of copper and tan. His chest feathers overlap in neat concentric ovals, like the paper-thin edges of a wasps' nest. These distinct colours and patterns indicate he is old, at least three years, maybe more. A mature musket is an exceptionally rare creature. Only a handful of English falconers have ever seen one in the flesh, much less flown and released one.

In the wild the little musket would be killed by a wide range of winged and four-legged predators. Consequently, he should be hard-wired to be highly fearful. On my return home, when he is removed from his travel box he displays a momentary wariness before quickly standing on my glove with surprising self-contained confidence. For an hour or more he remains calm as I walk about the cottage or sit on the sofa with the dogs. His behaviour transcends that of every other male sparrowhawk I have ever owned. In fact, he is the calmest of either gender I have experienced, a fantastic anomaly in the world of sparrowhawks.

When turning him on the glove, I notice that a large proportion of his tail is smashed to broken stumps. I will also discover that he has never been allowed to fly free or hunted. Until meeting me his life had been curtailed and contained by an existence spent behind bars.

Normally, a hawk has twelve perfect tail feathers. On closer inspection, two of this one's outer tail feathers have their tips missing. Three quarters up, five of the inner feathers are snapped in half. His ability to twist and turn in flight is seriously diminished: his damaged tail is a large enough disability for all quarry to escape him. Very much like Girl, and even if he was the most proficient hunter in the world, in his present condition his starvation and slow death in the wild are certain.

A fat and healthy hawk continuously regenerates broken feathers. A captive hawk does not have this option and new feathers can take up to six months to grow back. To circumvent this issue, falconers have developed methods to keep their hawks freshly feathered. Imping a new feather is a delicate process. A hawk needs to be held firmly without being suffocated or squashed. I gently fold my hand around the top of this musket's legs and the bottom half of his wings. He tries to wriggle, then relaxes. I drape a tea towel around his shoulders, roll him up and wrap several turns of masking tape around the outside. I rest the feathery parcel chest down on a pillow and his little flat head pokes out one end and he turns to stare while I cut and discard the snapped ends of his old tail. Lined along the desk are seven freshly trimmed sparrowhawk feathers. In the hollow centre of each, a smoothed and sandpapered strip of soft plastic is glued in place. I pick up the first replacement feather, dip the plastic tip in glue and 'imp' it into the cut shaft of his tail. Working from left to right, tiling the seven feathers like slates on a roof, I twist to fit, layering one after another into position. When finished, I pinch up and sprinkle a small amount of chalk dust over the join. This stops excess glue seeping out and sticking his feathers together. While the glue sets I attach new anklets, jesses, a bell and a leash. With the hawk tied securely to my glove, I cut the masking tape and the discarded cloak of tea towel falls away. He briefly flaps, whips back around and stands resplendent on the glove. He ruffles his whole body, looks about, twists his neck over his shoulders and fans his new tail up towards his head. Nibbling, he delicately uses waterproofing oil from a sebaceous gland at the base of his tail, smearing it up the shaft of his feathers. He does this several times, pinging the end out of his beak each time. With skill beyond human competence, each feather is now arrow straight and glossed to perfection. I marvel at his ability to decide when his tail is finished. How does he know it is correct? Maybe he can feel his feathers like a tortoise does its shell or a human does its hair. I take him outside and place him on a low perch in a sunny part of the garden. From the perch he jumps straight into his bath. The liquid rattle of his submerged bell sounds as I tentatively walk towards Girl's aviary. As I get to the door of her mews his name comes to me in a flash of logic.

I name him Boy.

At the side of Girl's aviary, I slide the feed hatch open and catch a sudden flash of wing. There is another momentary blur, then a soft thudding from inside. I can see Girl's feet through the small spy-hole. I count eight talons. A full set. Slipping on my glove, I step inside. The smell of a hawk ready

to be released permeates the air, a whiffy salted-caramel taste
of moist meat and carrion. I breathe in deeply, taste yeast and
the umami scoop of Marmite on chewy bread. Girl bounces
from perch to perch and dust swirls in shafts of the sunlight
streaming through the window. She hits my shoulder, flies up
into the netting then falls at my feet. I reach down to grab her
firmly. After eight months in seclusion any bond we had has
long since dissipated. She is highly fearful and aggressive. In
my hands her beak opens, she breathes heavily, her muscles
twang taut, strong, flexed, and she tries to wriggle free. I
struggle to tie her to my glove. When we step outside the
sun hits her body and she opens up her wings defensively,
puffs up her feathers, spreads her tail, doubles, triples in size,
and hisses like a serpent. She repeatedly pumps and powers
down on the glove, trying to inflict as much pain as possible.
She went into her aviary broken and immature but has now
undergone a startling change. As well as two new talons, she
is resplendent in full adult plumage. No longer muted brown,
or indeed the usual sparrowhawk slate grey, Girl is strangely
pale and her feathers glow in off-white silvery blue, as if
covered in a fine film of chalk powder. Her eyes are peculiar.
Offset in a weird combination, one a deep butter yellow, the
other a flooded luminescent fire orange. As a whole entity,
Girl is magnificently skewed, a pale ghost, a dangerous devil
in miniature.

I inspect her talons. The surface of each is bluish grey to

Bakelite black. Thin ridges – growth lines – extend from the root of each toe and emerge straight and parallel to one another. I see no cross-cut stress cracks, marks or weakness. The back one is the right length but has a bulbous lump on the tip. The other is a fraction of an inch shorter than it should be. Both are blunt. No matter, the curve of each new talon is correct. They are easily able to trap feathers. Girl can hunt.

In the kitchen of the cottage I cast her up, using the same method as I did when imping Boy. Using a thin file, I sharpen each new talon to a serviceable point and gently snip off the old leather anklets. When trying to refit new ones, I find the pattern is too small. If attached, they will lift scales, rub her legs and cause blisters which, if they become infected, will mean another trip to the vet. I fetch a new sheet of leather from the tool shed. On my return, Girl has escaped, slipping free of her casting like wet soap springing through fingers. I panic. I hope to God I have closed the windows. I find her perched on the door of the bathroom, glaring down at my intrusion. I chase her through the cottage and she flies towards the light and hits a window. I wince. She could have broken her neck. When I have reclaimed and bound her once more, I find the new anklets fit perfectly. I cut and punch two new leather jesses and push them through the brass holes on the back of her anklets. Without warning, Girl grabs both the leather and my hand. The flesh of my thumb rips on the tip of the talon I have just sharpened. I pull away. She struggles violently, is furious.

I hold her down with my uninjured hand. Each time I reach forward or move Girl's feet twitch and clap like the claws of a crab. Her speed and accuracy are unearthly. Over the next ten minutes she manages to foot me a further three times. When secured by her leash on my glove and at last released, she launches at my face, pendulums through the air, slicing, scything, grabs her own tail and spins insanely off the glove. Girl may be astonishingly beautiful, but she is also a little bit scary.

It takes fifteen minutes for her to calm down. I move with slow steps, like a broken, decrepit falconer, inching up through the garden as if I suffered from a muscle-wasting disease. Any faster and Girl resumes her swinging fighting stance. Fifty feet away, one foot tucked up under his chest, relaxed and calm, Boy watches us intently.

When we reach her designated area of the lawn, Girl refuses to be placed on her perch and repeatedly and ferociously leaps and bates to the floor. Two long trenches are quickly carved in the lawn. For a wild hawk, behaviour like this is fairly normal, but it can be tolerated only for a short time. Usually, a hawk will calm itself down, but Girl seems to know no such rules. The degree of her aggressive fear is beyond safe. She is relentless. Left any longer, she could break a leg or snap her feathers. The changes of the last few hours have been too much, leaving her on the lawn in this state will lead only to disaster.

I spend the rest of the day converting her aviary, constructing an impromptu indoor perch fixed to a shelf six feet off the ground. Inside the mews she will be protectively surrounded on all sides by solid walls, will feel less threatened, less vulnerable and less exposed than out on the lawn. From now on, when I approach her I will be at eye level and not tower over her. I screw a mesh screen over the doorway, which provides her with an interesting view on to the garden while keeping her safe if she snaps her leash or jesses. By the time I've finished it is getting dark and I transfer her to the shelf. She takes to it immediately, remains standing still on her perch. I close the mesh screen and bolt the door.

I bring Boy indoors and put him his perch next to the fire, slump down on the sofa and look at him. The inadvertent duality of the situation is not lost on me. Two wild hawks are ready at once. Barring further accidents and illness, Boy and Girl will be trained, hunted and released back into the wild world at the same time.

I watch as my son travels easily through the world of technology. He skips between tablets, phones and laptops, able to navigate and glide through Netflix, Amazon and a multitude of other online channels. His passion and interest are tangible. His ability to multitask, watch YouTube tutorials, play online games and hold a coherent conversation all at once is remarkable. He takes great pleasure showing me the details and tricks discovered in the invented inner space of Minecraft and other digital worlds of fun and fantasy.

I used to be passionate about technology. I am of the generation that has lived through the transition from analogue to digital. Long before the Web became a behemoth information-gathering device and long before powerful people 'moved fast and broke things', I believed technology to be a panacea, so much so I wrote an MA thesis on cybernetics, cyber culture and identity. As our reliance on technology has grown, I have become cynical and ambivalent about its sly intrusion into our lives. As the poet Gil Scott-Heron said about our greatest technological advance, space travel: 'Oh that's just whitey on the moon.' When pioneering new frontiers, it was ever thus. My choices and thoughts on technology are of no consequence to my son. He regards the whole panoply of gadgets the same way he does breathing: perfectly natural.

He invites me to join him as an avatar, a companion inside his games consoles, playing and participating in beautifully designed adventures. When I was a child, I did the same. I escaped into fantasy worlds, collected and painted small lead figurines, drew comic characters, invented my own mazes, spaces and role-playing games. Our joint quests are familiar territory.

For hours, we fight our way through synthetic, colourful, quick-fire levels, solving quests, unlocking clues and beating multiple overlords and enemies. The slick pull of these platforms and his absorption in their puzzles and complicated rituals are wonderful. He sucks up the information without a second thought.

As I am inside these games with him, a slow truth starts to emerge. Inside a screen with my son, walking around as characters, as a team, I find myself joined in a new world in equity with him. Free from the complexities and the pressure I place on myself to be a good father, free from the guilt and fear, the divide I feel temporarily dissolves. We work together, he helps me rediscover myself, I become a child again and he inadvertently reintroduces me to what first attracted me to technology all those years ago. He is training me to play, to extend myself and connect, and in so doing we build and shape new experiences through his love of technology.

Training

Fear, for a hawk, is the key emotion that enables survival. Most resident indigenous sparrowhawks have regular patrol routes, set routines and a detailed understanding of their world. Any fluctuation in their daily pattern is considered a threat, out of place, and naturally avoided. This heightened sense of fear is a simple, effective aid to longevity. Humans trigger extreme levels of fear in a sparrowhawk. Overcoming this fear using food rewards is the first step in training. Daily rations are marginally reduced until fear recedes enough that the hawk feeds from a small offering on a gloved hand. This reward for trust forms the basis of a positive association between hawk and human.

When they are taken to new locations, unexpected sights or sounds trigger new, elaborate levels of fear. Another slight reduction in rations is normally required until the hawk once again begins to feed confidently. The hawk is then attached to a thin line (a creance) and called from a perch to the gloved hand for a reward. The distance between the hawk, the human and the reward is increased incrementally. When the hawk flies a hundred yards or more instantly, the line is removed and free flight begins. Once free, the hawk will naturally become interested in the movement of prey and hunting can begin. How long this process takes is wholly contingent on the attitude of the hawk and the skills of the falconer in deciphering the moods and reactions displayed at each stage.

I have experienced a wide array of sparrowhawk behaviour. Some were calm, some mildly aggressive; others were frightened of post boxes, buses or deflated balloons caught in a hedge. Some bathed openly in streams; others only in private. Some hunted and killed easily; others took their time. Some preferred a particular quarry; others were adept at killing a wide array of species. There is no definitive pattern to a hawk's reactions when training. Each is unique, an entity with its own internal logic which takes time to unpick and solve.

Boy proves to be an exception. From the outset he is happy to feed on the glove in and around the cottage. On the first day of serious training I pick him up off his perch and place a whole chick in the palm of my glove. With the precise movements of a brain surgeon, he tips forward and pulls and plumes delicately at the soft coating until a small patch of fresh flesh is revealed. With two fast tears he is deep inside, feeding and focused. I call the dogs over and we start walking. As we move further out into the landscape, Boy continues to feed, unconcerned. By the time he has finished we are at least a mile from the cottage. He searches about for any last scraps and morsels. Finding none, he scrapes the edges of his beak along the top of my glove and ruffles his feathers. On the walk back, and without the distraction of food, I expect him to bate from the glove, to find an excuse to react with fear. He does not. We achieve in an hour what would usually take a week or more with any other hawk. His ability to glide through this stage is remarkable.

Before her accident and illness Girl was biddable and progressing in a fairly consistent manner. Now she will not countenance feeding on the glove inside the cottage or out. Slowly reducing her weight makes no real difference. In the back of my mind I know that the experience of her broken talon and lengthy seclusion has settled inside her body like a muscle memory. Girl displays deeply disturbed, almost schizophrenic characteristics. I find her almost impossible to 'read' or predict. She is a perplexing, confounding rule-breaker, both contrary and complex. She is acutely resistant to trust and now seemingly unable to form a lasting association between me and her reward. Occasionally, she allows herself to be drawn towards me but seems to catch herself and responds with abject resistance and protest. I watch as she lowers her head and feeds, momentarily relaxes, all meek and fearless, then inexplicably puffs up her feathers, leaps forward and hangs upside down under the glove. Her beak opens, and her feet snatch and spring for my hand as I reach in to place her gently back on the glove.

I think through the issue logically and come to the conclusion that the circumstances of her negative beginnings, falling from her nest, her many handlers and, finally, serious injury have been processed into a potentially unbreakable negative association between humans and distress.

I understand her nature.

My father has some exceptional qualities. He does his best, given his own particular upbringing, but remains a complex man. His enduring weakness is volatile anger. He could be fast with his hands, with threats and insults and has not spoken to any of his children for years. For us, the virus of aggression threaded through our lives disguised as loving actions, creating an atmosphere where verbal and psychological tension spread like cancer inside our young, sensitive minds. For me, it resurfaces, manifest in nerves, anxiety and internalized anger. For my brother, it took the form of running away, stealing, petty crime and short spells in prison. My sister has worked through her particular pain and is an exceptional and protective mother. In contrast, my brother has rejected his children completely. I fall somewhere between the two.

My grandfather shared similar traits and characteristics with my father. In this sense, generational anger was passed on like a baton, negative learned cycles of behaviour, rolling and repeating as the psychological trips and traps of parent-hood open and close. The deep-seated potency of this type of nurture is a twisted snare, making it almost impossible for any of us to escape unscathed. Most men fight with their fathers. I fight to leave this particular aspect of his emotional legacy behind. The biggest fear I have – my greatest fear as a father – is the threat of aggression arriving as a kneejerk, unconscious reaction to my son's behaviour.

The sun beats down, searing hard, as my son and I play on a trampoline. The heat on our feet and bodies is magnified by the black rubber. We sweat and fight and roll. He likes this game of rough and tumble and is remarkably tough, like a tiny cage-fighter. I throw him skywards, he falls back through the air and lands splayed out, bounces up laughing so hard he dribbles. From out of nowhere he catches me with a clear, powerful, left-hand punch to the eye. It hurts and I have to stop.

We continue to bounce and do battle. He is relentless, and the rules we have agreed on begin to slip, the more excited we become.

He suddenly bites me hard on the back, leaving teeth marks and a bruise. I flash pure anger, a white-hot rage. I have never shouted at or struck my son before, and I do not do so this time. It is very close. I hold myself back. I have had enough. I climb off the trampoline and go indoors. His mother takes over and I calm down.

The distraction works. A space opens where I can explain how I feel. It was my fault for getting you overexcited. He is sorry he bit me. I know in this moment of calm communication that if I had gone too far, crossed the line, there would be no way back, no space left for my son to return trust.

We agree to take the dogs out.

On the top of the hill a dozen skylarks are high overhead. They drop and ascend with beauty and skill. I point them out to my son and we watch in silence. He slowly moves ahead of me to a distance of two hundred yards or more and walks alone. After a while he asks to be lifted on my shoulders. I kick up dust and the dogs weave and twist between the apple orchard we are walking through. He clamps his hands under my chin and the bones of his bum hurt my shoulders.

In the silence, my mind wanders and I think it would be wonderful to walk like this across Europe, through the Middle East and up through Mongolia and Russia. Cross the Bering Strait from East Cape to Prince of Wales, walk down through North America into Mexico, arriving in Costa Rica in time to watch turtles lay eggs on his eighteenth birthday. I tell him about my dreamed journey:

'Don't be silly, I have school on Monday.'
I tell him we could just bunk school.
He considers it for a while: 'OK, but only if Mummy can come.'
I start making up silly rhymes about the names of countries and doing funny voices, impersonations of the dialects we would hear along the way. He laughs hysterically. I watch a thin line of dribble slip to the ground in front of my face as we walk home.

Walking Girl to the training ground takes an inordinate amount of patience and effort. She looks for any excuse to become startled and throw herself from the glove, with or without a reward. I have to hold back my own frustrations. When I eventually tie her to the creance she flies to the fist, but only up to a certain distance. If I move further than thirty feet, she hits the glove, snatches her reward and flies off, landing on the ground some distance from me. Normally, this would indicate one of two things: either, she is overweight and not focused on the food, or she is simply scared. I know from watching her eat, from her body shape and posture, that she is at the right weight. To lower her any more would be unwise. I give serious consideration to waking her, to push her beyond fear through the lens of a trance, but I have no one on hand who can help. Instead, I take a longer, circuitous route and begin spending all my time with her, carrying her around inside and out for the whole day. To keep her occupied, I let her pull and tug on the fleshless bones of an old pheasant. This begins to work, but only to a degree. The behaviour she displays cannot be adjusted with negative responses. I know the futility of shouting at her. I am unable physically to force her to perform. In an attempt to remedy the problem, to remove myself from the equation and lower her levels of intense fear, instead of calling her to the glove I need to put distance between myself and her reward. My plan is a simple one. I attach the chick to a padded leather lure and swing it out on a long string on to the ground in front of her.

On the first day of trying, Girl looks at the lure lying in the grass and remains resolutely on her perch. Distracted, moaning and complaining to Flash and Etta, I briefly turn my back. Girl moves fast and silent and I do not hear her bell. She hits the lure with tremendous velocity. The creance cord whips off the reel and spins across the floor, and Girl takes off over the field with the lure, her full rations and the creance. As she attempts to land in a tree, I watch a section of the creance catch on a branch, and Girl spirals around and around midway up the trunk. It's too high to climb, so she hangs struggling and flapping for the half-hour it takes to co-opt a friendly farmer's JCB digger and be lifted up in the bucket. Once untangled and brought back to earth, she is incandescent with rage. It is a total disaster. It takes two days before Girl even remotely trusts me again.

~

Boy is a long way further forward and ready to fly short distances to the glove. I place him on a post, turn and walk away. At fifteen feet I fumble for the reward in my right pocket. I hear the sound of his bell, feel a light, soft gust of wind and a sudden pressure on my head. Boy has flown to me without a reward and landed on my hat. I begin to giggle and reach up to grab the thin cord keeping him secure. It is not there. I turn and see it is unattached, snaking through the grass up to the post. More by accident than design, Boy achieves free flight under his own volition. The feeling is reminiscent of narrowly avoiding a car crash. A sickening mixture at the terror between what *could have* happened (his loss) coupled with the explosive joy at what *actually* happened (his safe recall). I wait a few minutes for my heart rate to reduce, then return Boy to his post. This time I face him and walk backwards to roughly eighty yards. I am pushing him, doubling, tripling, a distance reached only after several days of flying on the creance. A large tractor passes along the lane, heavily laden with bales. Too far away to reach him in time, I freeze. If startled, he will take flight, and I need to be ready to run after him. The roaring noise, smashing branches and swaying load of stacked straw hits trees as it moves along the lane. Alert and bobbing his head, Boy's focus remains directed entirely on my actions. I hold a chick leg in my glove and raise my arm. The speed of movement, the tight thrum of his wingbeat and the brief time it takes to cover the distance are astonishing. Once the leg is eaten, I sit with him in the grass and let him eat his full rations. He drops tiny morsels to the floor, accidentally rewarding the dogs. They come in quickly, shunting and sniffing the floor like steam trains, lick their lips, then my face.

Boy continues to relish the feel of flying free. He returns consistently from trees and long distances like a sublime feathered dart. The added exercise triggers his rapid-fire metabolism and the little hawk's deeper instincts begin to emerge. He shows an acute interest in anything that moves, then makes contact with his first wild animal.

Walking the uncut edges of a hayfield, Boy spins from the glove, intent on catching the invisible. Forty feet out, he flips over and disappears into the long grass. A bucking bronco lifts up out of the cover and my heart stops dead. Boy is attached to the back of a full-grown rabbit. No more moved than if a fly were attached to its fur, the rabbit careers towards the bottom of a fence. For a six-ounce hawk to try to kill a four-pound rabbit is utterly wrong. In this split second I know Boy has never hunted. He is a blank slate, he has no idea what the correct quarry is. The rabbit reaches top speed and zigzags back and forth towards a tightly strung wire. If he hits it, Boy will be killed instantly. He adjusts his grip, lets go and glides up on to a post, touches the top, pauses then volleys back to my raised glove. His hedonism is beyond question. It was close, far too close. There is nothing more to learn today, no more risks to take. I feed him his full quota and we head home.

Crossing a little bridge near the cottage, Boy becomes transfixed by the reflected light of the stream running underneath. He bates towards it, so I let him go and follow. He lands gently on a wedge of sand, walks over a small patch of gravel, dips down, sips then steps in. The late-afternoon sun splinters the canopy in shafts of mottled auburns and pale yellows. Petals of white hawthorn fall free from their branches, cut across the warm breeze and land on the water like the tipped contents of a hole punch. Boy vigorously begins his first wild bath. Exposed and with wet feathers, a bathing hawk makes an easy target for predators. Confident in my company, this rarely seen, secretive behaviour is a joy to witness. Photographers and film makers could wait a lifetime to see it. The vaulted spray from his feathers scatters, arches over in tiny droplets and tings my skin. I reach down to throw water on his back. His vibrations and contortions become even more frenzied. Scooping up a handful of water, I raise it over his head. He follows my finger tips and opens his beak. I trickle water into Boy's mouth and he drinks deeply and swallows. When he's finished he walks up on to the sand and wriggles off the excess water like a tiny, wet dog. I hold my glove towards his chest, he hops over the ground and steps up.

The water has parted his feathers and his crop, as round as a golf ball, is tissue thin with threading capillaries over a taut, flesh-coloured surface. On the side of his head his ears appear as two holes drilled and disappearing into his skull. In his drenched state, Boy is the oddest, ugliest and most beautiful little hawk in the world.

I generally dislike the feel of water. It makes me itch, like having cockroaches stapled under my flesh. I only go in water when I feel there is good reason. My son likes swimming and wants me to take him. This is the first time I have taken him swimming on my own. When we get into the pool he displays total freedom. I, on the other hand, am deeply self-conscious and embarrassed. I hate it. The noise and cloying heat is painful, loud, brash. The warm, chlorine atmosphere is a drenched chemical overload and faintly suffocating. My senses are bombarded by the bright, bouncing strip lights. The close proximity of other semi-naked humans is viciously uncomfortable. My son does not give two hoots. He bumps into people as he somersaults backwards and forwards off the edge of the pool. It is lovely to watch him be so free. I feel the freedom of his physicality, see it in the small stretches of his muscles. The water washes over his legs and torso like fresh rainwater. Under the water, bubbles roll and fall through blue and green foam and catch in his hair and in the corner of his eyes. He is a pale and perfectly formed fish. He climbs out of the pool with ecstatic smiles, his brown eyes blown wide and bright. He shuffles up and down on his feet in excitement. He jumps on to me and we slide and roll and bounce and glide and move together like otters. He goes on the water slide; we play hide and seek. We float about on large pieces of foam. Time drip-drips away and only then do I realize how long we have been in the pool.

When we get changed it is the first time he has seen me naked. He laughs at my willy, says it looks like an old nose in wire wool. I laugh. I give him a Jaffa cake to shut him up. On the way home he falls asleep in the car and purrs like a cat with its mouth open. He is so beautiful I nearly crash the car, watching him sleep.

When I finally remove the creance from Girl we fail to achieve any level of consistency. Her reactions often beggar belief. She changes expression and attitude hourly and will fly three or four hundred yards for a tiny chick leg. Ten minutes later, a whole quail carcass tied to a lure pad and thrown forty feet below her will scare her up through the branches into the tops of trees. She is less a sublime feathered dart, more a badly designed shuttlecock. She remains unlike any other hawk I have owned, trained or seen flown, in England or abroad. Girl is an enigma. She has no inclination towards and little regard for humans. She is the worst falconry hawk I have ever trained. As a hawk to be released, however, one needing to survive on evolved wit and instinct, her belligerence is pitch-perfect.

But her stubborn fear forces me to the outer edges of my experience and we arrive at an impasse. I have to admit a kind of defeat. Rather than hold her back, happy that she will haphazardly and violently return fifty per cent of the time, we need to move forwards. We need to catch up with Boy.

Entering

In the wild, sparrowhawks lay, incubate and hatch their young a few weeks after songbird chicks have fledged. Wobbly and unsure of the world, these young birds make easy targets and a regular source of protein for a nest full of rapacious raptors. When feeding their chicks, adult sparrowhawks will bring a wide array of whole carcasses and wounded live prey for them to consume and 'play' with. From the moment a sparrowhawk chick breaks its way out of an egg it associates different shapes, movement, textures, tones and tastes with food.

When a sparrowhawk chick has left the nest, is fledged and flying free, its prey grows in matched proficiency. They share an equilibrium, both learning different methods and modes of flight, one of attack and one of escape. The hawk, through repetition in success and failure, learns which quarry is weakest, which is toughest and which gives the biggest meal. A healthy hawk expands psychologically, grows and develops with every success as the season changes. If I cannot match this process exactly, if I inhibit this delicate part of their rehabilitation, it will cost Boy and Girl their lives.

My first sparrowhawk, Daisy, was a pale, slender, feathered barb of a hawk. A few months before she arrived, I reared and released a large number of red-legged partridge. Instead of working the land in a natural manner, it was easier to wander the same fields each day and flush a partridge. Daisy caught them easily and we both enjoyed eating the fresh supply of free-ranging meat. As the seasons changed she built an association between herself and the partridge so powerful that she refused to hunt anything else or indeed make any effort to fly anywhere other than the fields in which they lived. Naïve, and without realizing it was possible, I arrested her development, narrowed her experiences and mind-set down to one species and one location. I failed to understand how and why a sparrowhawk has evolved, how closely they are connected to a diverse range of quarry. Instead of developing her abilities in an expansive, natural way, I sealed off a huge part of her existence. I melded her to what I wanted and not what she resolutely required. She was what falconers call 'wedded' to partridge. If lost, she would have been entirely incapable of surviving on her own. The same principle now applies to Girl and Boy. Killing and hunting as if they were free and wild is a critical component in their race for life when free. They have to be able to compete with a wide range of different quarry and win.

—

My son is on the starting line in a race. Set along the one-hundred-metre lanes are various items of clothing the children have to put on before proceeding to the finish line. The flag goes down and my son runs like the wind. On the first and second stages of dressing he is well ahead. For reasons I cannot fathom, I scream his name as encouragement. It has the opposite effect. My voice completely short-circuits his concentration. He hears me, stops running and looks at me with a puzzled expression. All the other children run past him. I am devastated. My presence has destroyed his chances of victory. After the race he agrees that it would have been best if I had kept my mouth shut. He is magnanimous in defeat, neither disappointed nor upset. This is the first sports day I have attended and he seems more delighted that his friends can see that his father is here to watch him run in the races, to compete, to win, or in this case make him lose.

My turn comes around too quickly. The tannoy crackles and announces: 'Those wishing to attend the fathers' race, please make your way to the starting line.' I am hyper, my heart is beating fast and I stand on the starting line long before any other competitors. When all the other fathers are lined up I cannot help myself. I start making highly inappropriate (and untruthful) jokes about running away from the 'the filth' for theft and burglary. They do not laugh, they just look at me. I am failing to make a good impression. When the race starts, I also fail to come first.

Thankfully, my son is too busy with his friends to notice. His mother, on the other hand, filmed it all for them to laugh at later. I do not care. I gave it my best shot.

For four days running Girl and Boy fail to catch or kill. The heat is exhausting, the summer cover high and impenetrable. Moreover, the natural world is now properly attuned to the presence of a hawk and its animals become super-sensitive in order to survive. The world within eyesight reinvents itself and different types of behaviour not normally seen are triggered. As we walk, within a roughly 500-yard circumference there is total silence. It is like being inside a bubble. Birds and mammals remain hidden, sit tight and quiet for surprising lengths of time. In the past, such is this fearful stillness, I have been able to reach into bushes and pick birds out by hand. Outside this silent surround, the noise is one of alarm. *Tic-tic-tic* whines and the screaming chatter of warning spills out continuously. Swifts and swallows will dart and bomb the head of a released hawk. I have had wild sparrowhawks attempt to swipe a falconry hawk off a perch in the garden; others were chased around trees or grabbed off the glove. All this defensive commotion, the fluctuating behaviour patterns and natural avoidance work against the hunting efforts of Boy and Girl.

With each failure Boy shows infinitesimally small levels

of progress. To the untrained eye, it looks like he is repeating the same mistakes. I see slight recalibrations, an awareness of new tricks, and instincts laid dormant begin to emerge. His wing beat changes, as does his posture. His speed of reaction increases, the corners he turns become more acute. The areas he selects for a springing, sprinting chase are well away from protective cover. He eases into trying out all manner of slips. Some short and stabbing, others, long, quick, pulsing pursuits. He begins zigzagging over ditches, down along hedges or up and over on a sneaky surprise. Slowly he becomes his true self, flying and hunting like a line of white light reflected off a spinning disco ball. On numerous occasions, if he misses he boomerangs back through the air and lands on my glove, ready to go again. At other times he crashes bravely into the cover or pulls up over where the quarry has landed. By marking their point of entry he goads me to re-flush, to go again, and again, and again. He gets so close that at times the difference between catching or not is no more than a blade of grass, a twig or a leaf getting in the way. We just need a bit of luck, some slight weakness in the quarry, a missing feather, illness, or a shaft of sunlight glaring into the eye of his chosen targets.

On the fifth day I rise at dawn, weigh him and, as the sun breaks, we set off. Some mornings feel right, a day when a sixth sense kicks in. Not in a mystical way – there is nothing supernatural or spectral in the moment. I make no incantations or promises to invisible gods, but years of flying hawks have attuned me to a very specific state of mind. The feeling remains the same. It never wavers, changes intensity or dips, and I have never felt it in any other sphere of my life.

As we progress, the little details collapse into focus and join together. Every aspect is a one per cent click towards success. I see animals in unexpected places, further out than usual, exposed. The light is right. My feet fall without sound. When we change direction the wind moves with us, always into our faces. The temperature remains cool. The hawk is keener. Boy buzzes on the glove. Etta and Flash move differently, focused, dancing through scent with grace and style. This feeling mounts layer upon layer, overlapping and interweaving. A hundred yards out, Etta shines copper coloured against iridescent green, turns right then cuts across the breeze and stops dead. Her muscles tense, trapped along her spine, chest and legs. Pheromones wash through her nose in an organic chemical overload. Her tail sticks straight out, a dog-shaped statue, marking and pointing. This is it. The world sucks together into a silent, flowing, continuous whole. I adjust Boy's jesses and reposition him, he twitches, shuffles then grips the glove. My mind unhooks from spinning wheels, from fragments and thought. The light and my eyesight ramp up brighter, clearer. My other senses descend and fall away. I hear no sound, smell nothing, feel no sensation, no cold, or wind. I am out of my mind and absolutely present.

Etta edges forward then stops robotically. I stop. She moves. I move. Flash runs in. Two birds shoot skywards ten feet and turn. One slows and skims parallel to the ground. Boy flies the first, checks, reselects, snatches, catching the second bird in mid-air, and keeps moving.

I work towards the sound of Boy's bell. Breaking through blackthorn, I find him tucked tightly under a gnarled root, removing feathers from the wings of his kill. The rush of adrenaline and my joy for Boy roll over me in immense gratitude. I force myself forward into the small space, spiking my hands and scratching my face, and lie prostrate at his feet. I reach out, touch his kill and thank it. Boy stops eating, looks directly at me, then carries on delicately stripping feathers. With this small gesture of trust, I feel no separation or distance between myself and the hawk. I have been welcomed inside the electrifying loop where Boy is at the end and the beginning, his life rising as another descends. This little big bang, this tiny bag of feathers, has the gravitational pull of a planet. He has drawn together five separate life forms – a human, two dogs, a hawk and his quarry – across instinct and evolution. We have entered his world, are actively participating in a highly condensed, private moment of survival. It is a privileged position to be in and never loses significance.

On the walk home Etta and Flash trot ahead and Boy is secure on the glove. Small feathers are stuck to his feet and legs. Now he has eaten everything, his head is pushed back and he cannot see his feet. I detect a definite change. I see a definite change. Boy is coiled down with new knowledge, the final piece of his internal jigsaw pushed into place. He is ready to flick out and feast on the whole world. He lifts his left foot up and tucks it into his chest feathers. As my adrenaline drops, my senses return, sweat cools and the pain arrives. Using my teeth, I pick blackthorn tips out of my arm. Several remain in the balls and palms of both hands. In a few days the poison will bubble and turn my skin a bruised blue and the infection will weep pus. Nearing the cottage, my extreme elation sours to sadness. Boy's journey back to the wild has begun in earnest. This first step to a new world of freedom takes us closer to separation.

Bored with computer games, I tell my son of a world with mythical beasts where strange characters exist, one with elemental challenges, puzzles and adventure. For us to succeed it will require the dexterity, focus and hand–eye coordination of a gun-slinger. I explain that the journey, and its prizes, are not without risk or failure. That once he accepts the challenge, once our journey begins, he will enter into extreme levels of vertiginous fear, joy, possible sadness and ecstasy. He doesn't believe me at first and laughs. Tells me I am talking rubbish and asks for a biscuit. I show him some YouTube clips. He is instantly hooked. So we set sail in preparation.

On the morning of our expedition we collect supplies. I give him a shopping basket and let him forage freely along the aisles. We will be travelling fifteen minutes from home and be gone for no more than two hours but, in case of an emergency, he has collected twenty-four sausage rolls, a tub of trifle, some chocolate mousse, a strawberry cheesecake, ten packets of Wotsits, Kinder eggs, lemonade, apple juice, strawberry bootlaces, roast chicken, Hot Wheels Slime, a scratch card (his first) and a lottery ticket (also his first).

By the time we arrive it is a blindingly hot summer's day, a day when you sweat in the shade. We crunch down a gravel track, pass through a gate and stand on the edge of a lake. This is the first time I have taken my son fishing. We place our tackle on the ground and take a walk. Like all good fishing lakes, this one oozes ethereal potential. The bankside vegetation is thick and lush. As if by magic, a grass snake materializes, swimming across a flat area of open water. We watch a water rat zip along the grass. Ducks and ducklings skitter across the surface. In a covered corner, the light breaks through alder trees and dapples off the surface. Pinprick bubbles fizz and pop along the outer edge of lily pads and the water is cloudy with clay particles. Fish are feeding, grubbing about on the lakebed. Huge, pale purple submarines cruise just below the surface, then drift down, disappearing through the water. Instinctively, my son's behaviour changes. He is pointing and talking in whispers. The only thing that would make this moment more magical is a unicorn.

Sneaking about like bone fide hunters, we come across an old man and his wife tucked behind a huge line of reeds. I quietly ask if he has caught anything. As he turns to talk, the rod in his hand buckles and he strikes into a big fish. The line zings through the water, parting the surface scum and old feathers like cheese wire through butter. Huge boils and swirls appear on the surface. Mud, black sediment and broken leaves rise up off the bottom. My son clasps my side in pleasant terror: 'What is it? How big is it? Is it a catfish? Will it bite? What is it? What is it?' The fisherman rises from his seat and with unfaltering generosity offers my son his rod, reel and the hooked fish. It is an act of supreme understanding. Every angler knows the importance of a child's first fish, how the moment becomes indelibly etched on to the mind and soul for ever. My son pauses, unsure. I hold my breath, don't say a word and look at him. Curiosity kills the cat. He takes the rod from the stranger and enters a whole new world. The fish bolts, the rod bends at an alarming rate, pulling my son's arms straight and hard against their sockets. His eyes dilate as wide as dinner plates. He does a little dance in the dust. His mouth opens in the shape of an 'O' but no sound comes out. He holds firm. The whizzing, clicking buzz as line peels from the reel is the only noise we hear. The water rocks up and a bow wave surges out towards the island. Whatever is on the end is royally duffing up my son. Scared, he wants to give the rod back. Both the fisherman and I decline. Instead, we start shouting snippets of conflicting advice: 'Turn it left, no right, reel forward... no,

the other forward… reel… REEL… NO, STOP!! Let it run…
hold on… put the rod tip lower… watch the branches.' For a few
glorious minutes we are totally out of control. The fish runs us
ragged, metaphorically tweaking our noses and kicking us in the
pants in the most delightful way possible. Eventually, the fish tires,
turns and rolls on the surface. The man scoops it up in his net.

On the soft grass, the tench is as wide as my son's shoulders. It
weighs at least eight pounds, double the size of my own personal
best. It is a stunning male in deep dragon green, its olive flanks
contrasting with circular orange eyes the colour of a goshawk's.
Giggling, my son runs his fingers along its side and tells me the fish
feels smooth and slimy. I tell him a tench has magical, medicinal
properties. In the days of witches and druids it was called the
Doctor Fish, its slime a cure for illness and maladies. I ask if he
wants to lick it. He tells me not to be stupid.

We return the fish carefully, thank our new fishing friend and
head off to our own corner. Five hours later, we are covered in fish
slime, he has chocolate mousse around his mouth and in his hair.
We have caught a lot of beautiful fish. It is a day when my plans
and quiet expectations match reality. It is the first time I have felt
the deep joy of paternity.

Entering Girl into the world of wild quarry proves a lot harder.
Her inconsistency means she flounders and keeps missing.
When she misses, I find her empty footed, tightly bound to

twigs and rolling around like a disturbed child. She will not let me touch her until she is utterly calm. I wait up to fifteen minutes before she decides to clamber on to the glove with a footful of leaves and blackthorn twigs. She begins to collect multiple scratches, cuts, piercings and bumps on her feet. She does not take kindly to me spraying her feet with disinfectant and scrubbing them with a toothbrush each day.

Sexy Lexi was another female sparrowhawk I owned, and she would act very much like Girl. On nearly every miss she would foot branches with a powerful pumping movement. Half a dozen machine-gun pulsing clamps within a second. Perfect for killing birds but not for spiked branches. A blackthorn tip snapped off, embedded into the soft pad of her foot and a small scab grew over the top. The swelling, a build-up of pus, crystallized into a hard, flinty fragment the size of the rounded sulphurous tip of a matchstick. Lexi had to be sedated and undergo surgery at a cost of around £1,000. If Girl continues to behave in this manner, it is only a matter of time before she succumbs to a similar injury. The only solution is to keep away from blackthorn hedges and strike out into open fields. Here there is less chance of Girl further injuring her feet but also less chance of making a kill. It is a frustrating choice: the situations are as bad as one another.

The following day Girl is hot, cantankerous and frustrated. I match her mood perfectly. We have been walking cut, flat fields for three hours with no success. Absent-mindedly, I cross to a cluster of corn stalks untouched by combine harvesters. A bird breaks across the gap and bobs and tips about on the floor picking up discarded seeds. We all look at one another. Girl doesn't even bother to chase. The bird takes the opportunity to escape and flies to a small, frothy patch of nettles under an old aluminium water trough. The bird is easily bumped out by the dogs. Girl changes heart, grabs her meal ten feet out and eight feet up. Unexpected, sudden, and fast. I am stunned. Girl has done it.

The relief is brief. We move from poetry to pain in less than a second as I watch Girl fly hard and fast, curl up over a hedge and disappear into the landscape. For half an hour I sweep the sky with the telemetry. Nothing. Switching to another channel, I get a positive signal and track her down three fields away. I hear the smallest, muffled sound of a bell. I look on the ground, fold over patches of long grass and lift branches, trying to find her.

At the base of a large oak one or two finger-length feathers are stuck in the ground. The leaves above are dense and too thick to see through. Standing against the rough trunk, my arms wrapped around the circumference, I stumble in a circle looking up. Halfway out on a branch, Girl is eating her kill. Clever Girl! She has learnt to fly to elevated cover and to feed in safety. This is good for her survival, not so good for me. There is no way she is coming down until she has finished eating. She may also move again, if disturbed. Sitting on the ground directly below her position, I roll a cigarette and wait. Girl plucks and pulls at her meal. As if a fairy has burst a pillow, small feathers rotate down from the branches, touching and tickling my face. When the snowstorm stops I look up and wave her daily rations around in my hand. I am expecting her to refuse, but she drops from her branch and lands on my glove with a soft bump, then leans over and begins to eat my offering. I nearly fall over with shock; she is almost normal. The relief is sublime.

There are very few things more delightful than a wild hawk making its first kill. The laughter of a small boy being tickled until snot bubbles form is one; the other is the realization that halfway through a walk his wellingtons are pointing in the wrong direction to his feet. His boots are on the wrong way round. We cry with laughter as he marches about like a court jester then parades up and down, moonwalks and refuses to put them on the right way round.

Boy's aptitude and number of kills rises steadily. Girl's remains the same: one. Her failure is followed by a slow return to frustration and anger. I have no choice but to once again change tactics. Both methods I have in mind add subterfuge, both will provide Girl with a better chance of realizing her full potential. Unfortunately, both are inherently risky.

In the past, falconers in England would fly hawks and falcons from horseback to gain a height advantage and cover rough ground easily. In a modern context, the form is different, but the theory remains the same. Some falconers now use trucks and off-road vehicles for the same purpose. Crows, pigeons and magpies are less wary of a truck crossing farmland than of a falconer walking with a hawk. The element of surprise usually tips the outcome in favour of the hawk. Precluding the idiocy of human accidents, car hawking works well.

Years previously, driving in the fields close to my home, the hawk I was flying had caught and secured a magpie. I jumped from the truck and ran the short distance to help. Kneeling down, focused on watching her feed, a strange beeping noise broke the adrenaline rush. I turned around, confused, and watched in horror. The Toyota Hilux was moving steadily backwards across the path, the dog looking passively through the open door. When leaping from the automatic truck I had accidentally kicked it into reverse. Before I could act, over a ton of expensive truck (which I had borrowed from my son's mother before he was born and which, technically, did not belong to me) disappeared into a ditch, rolling towards a farm pond. The open door caught a tree and the truck slowed to the sound of creaking, squealing metal, followed by an exploding, bursting sound as the door bent backwards and the Hilux carried on down the slope towards the pond. The damage took a lot of explaining away (I blamed the dog) and I vowed that it would be my last foray into motorized falconry.

Aside from the use of a vehicle, the only other option is to let Girl fly free from trees like the two eagles in Germany. Unfortunately, after a handful of captures, Girl would be even more difficult to control. She would begin to actively search out high positions, become wholly reluctant to return and start self-hunting. In falconry terms, encouraging self-hunting is bad since the potential for losing a hawk is huge. For a rehabilitated hawk, it is perfect. The balancing act would be knowing when to stop. If I pushed too far, allowed her too much freedom and she failed to return, Girl would be flying free with a bell, anklets and telemetry. The loss of equipment is of no real consequence, but the noise of a permanently attached bell would draw attention to Girl for the rest of her life. In the summer, this is not an issue, as the innocence and inexperience of fledging youngsters tips the balance in Girl's favour. In heavy winter, her quarry hyper-alert, fit, robust and with less cover acting as camouflage, the bell would be easily detected. A bell left on her leg would mean the difference between survival or death.

If free flight were to work, I would have to control all the elements closely. It would have to be the final stage before release. All the planning in the world will not stop it from going wrong. If it fails, the love between hawk and human is easily replaced with frustrated disappointment and a directionless hatred of falconry.

I have always wanted to take my son to the seaside. The seaside in my mind is the one I love the best, the one I want to share with him. It is a place of happy childhood memories, and a location I return to as an adult when I want the taste and smell of the sea on my tongue. Situated on the Welsh coast, it is a small natural cove that doubles up as a tiny campsite for half a dozen people. The farmer allows fires and sells burgers and chops from the lambs he butchers from his back door. The sun sets perfectly. I have seen porpoises, caught huge sea bass in nets and pulled lobsters from deep fissures in the rocks. When the tide turns, the pools heave with creatures: a handful of smashed limpets or snails makes the floor move with life. Thumb-sized anemones begin to wave like palm trees in a gale. Crabs, an assortment of shrimp and prawns, club-headed gobies and bootlace conger eels scoot out from rocks and chomp down on the leathery discarded flesh. It is a place where buckets are filled easily, where time slides away and where you get sunburnt without realizing it.

Unfortunately, this place is about eight hours from where my son lives. A daytrip is out of the question. A compromise is reached. He lives forty-five minutes from the south coast. So we head off for the day, our expectations and planned adventures well beyond sky high.

At first, we cannot find anywhere to park. All the car parks are full. All the streets have double yellow lines, signs and traffic wardens in numbers almost equalling the crowds. When we reach the sea the tide is in and not due out for another six hours. I had forgotten to check the times. The water of the English Channel is a swirling, murky, mud-brown soup. There are no rock pools. The beach is stunningly stark, just a long line of shale and shingle. Admittedly, the stones in between the lumps of discarded plastic are beautiful, but the beach is uniform from left to right to the width of the horizon. Someone allows their dog to take a shit right next to us and fails to pick it up.

Undeterred, we start skimming stones, building piles of rocks and knocking them over. We make collections of the prettier shells, comparing their sizes, shapes and colours. We walk and explore. Etta and Flash swim and chase the seagulls eating chips from bins. We find several mermaid's purses and cuttlefish bones. If I put a positive spin on it, it is not too bad. After a couple of hours, we get an ice cream and drive home.

Halfway home my son turns to me with a serious face:

'Well, that was a SERIOUS disappointment.'

I roar with laughter. It is exactly what I was thinking. It is very honest. Totally inappropriate. The sort of thing I would say. He has no idea how happy he has just made me.

From nowhere, an unexpected wave of emotion washes over me, arrives without forethought or prompting. I am shocked by its natural power. I try to control it, force it back down, then give up and let it run its course. I turn and say:

'I love you.'

This is the first time I have told him. This is the first time he is old enough to listen.

On the first morning of free-flying I take Girl to a small wood near the cottage and sweep my arm sideways and up, letting go of her jesses. She lands in a tree above a thick blanket of nettles to my left. The resident blackbird, thrush and other woodland birds begin their clicking, ticking alarm calls. A pair of wood pigeon explode out of a thick mass of ivy on the trunk of a dead tree. Several smaller birds break from under my boots and Girl moves. I hear her bell above and I assume she has chosen to pursue. I pause to listen. If she was flying, or far away, the noise would recede with the distance. It doesn't. Instead, it remains clear and loud, chimes in a regular beat. Ting… *pause*… Ting… *pause*… Ting. I smile: this is a familiar marker. The jolt of flesh ripped piecemeal from a kill moves a bell's dapper with the same force, in the same direction, creating the same noise; Ting… *pause*… Ting … *pause*… Ting. Girl has killed. I look about, trying to pinpoint the direction. The sound is delicate and is thrown about like a ventriloquist's voice. On squeaking wings, the pigeons sweep back above me, touch the ivy, startle and immediately break out of the wood in different directions. I have found Girl, and Girl has found the pigeon's nest high above me.

I have no choice but to climb.

At the top of the tree I find Girl perched precariously on a strange mass of old twigs. She has a partially fledged dead pigeon in her feet. Its brother or sister looks on, unconcerned. Trying not to fall, I lower myself along the tree and crawl in deeper. Girl keeps feeding. I inch forward and hold the dead pigeon still. Girl adjusts her feet and her left foot gently spreads over the back of my hand. She continues ripping at the skin and soft meat, breaks into the chest cavity and slurps up warm blood and fresh liver. I am no more than four or five inches from her. I can taste the stale dust of dry, powdered excrement, see the curling unbinding skin on twigs and the whitewash slash of pigeon shit over the edge of the nest. I smell the blood and see the herds of red mites running over Girl's talons and across my hand. Small insects and beetles slip and slide between rotting feathers in the bottom of the nest. I hear flies buzzing and feel the waxy green leaves of the ivy touch my face. Damp moss wets my knees and chest. Splinters of rotten wood and dead ivy scratch up my sleeves and fall down between the cheeks of my arse. The pigeon which is still alive has a bulbous beak and deadpan eyes and shuffles back against the ivy, breathing steadily. It is a fairly grim situation but not without a silver lining. I figure that, with one sibling gone, the remaining chick will get double rations when we have left. Even if it doesn't, Girl has taught herself another incalculable lesson. A sneaky, opportunistic trick, of use when she is free. This is how Girl and Boy will hunt most of the time. This is how wild sparrowhawks survive, and pigeon is an excellent meal, the best. The kill took low levels of effort for a high-energy reward. For two or three days Girl's body will benefit; she will feel different, stronger, powerful. The association made will no doubt help keep her alive when free. I pull Girl up with the remains of her breakfast, tie her tightly to my glove, pat the remaining pigeon on the head and slide like a chimpanzee down the fallen trunk.

I am suspended high above the ground in a harness clipped to a thick metal wire. My son is in front of me and we are a little bit scared. We are on an assault course called Go Ape! in the woods near his home. There are a lot of other families with children, all of whom are at various stages of completing the course. It is a clever design and good fun. After we have done a couple of laps we begin to get a little bit cocky, a bit brave. We start rushing and challenging ourselves: one-handed, hopping, eyes closed, backwards, daring each other to do silly things. My son gets more and more excited and enthusiastic. We run up against the family in front. Their little boy is the same age as my son but a lot more hesitant. They are both going across a shaky bridge on a sky platform and I can see my son beginning a whispered conversation with the boy. He is quiet, the words just below audibility, a low mumbling that forces the other boy to keep turning around. The boy grimaces and he begins to get angry. I know exactly what is happening. Only a child with my genetic make-up could possibly begin to cause an argument suspended fifty feet above the ground. I know I shouldn't, but I feel a strange pride. I used to get into fights a lot in secondary school. I had a compulsive urge to say things, always to the wrong person, usually teachers, often bigger, older pupils. I have learnt to tame it as I've got older or to avoid situations where it is likely to happen and I can still see him whispering, like a naughty macaque. The other boy is now seriously agitated and starting to wobble about. Thankfully, his father is below, tapping away on his i-Phone; otherwise, it would no doubt have escalated into an intergenerational confrontation. I hear my son say, 'I am only being sarcastic.' That is not a good word to hear from a small boy. It generally spells trouble. I intervene just in time to stop it escalating into a full-on monkey fight. My son smiles mischievously, is coy. I wink at him. There is a silent recognition between us as we swing on then slide down the ropes, landing with a soft bump.

I continue to fly Girl from hedges, trees and fence posts. She has the most success working a long line of larch, catching young magpies, then a jackdaw. Realizing bigger birds are a legitimate food source, Girl takes to chasing crows, on one occasion for over two or three hundred yards. I know this to be dangerous. As she is so close to release, I try to curtail her enthusiasm, but she is vigorous in pursuit and actively searches them out. In her first year Lexi would do the same, with near-fatal results.

The single young crow selected was in the centre of a field and vulnerable. Not spotting the hawk until too late, it lifted in panic and Lexi rotated up underneath, snatching it twenty feet off the ground. They fell like two stones spinning on string. It was awkward for me to get to her quickly: I had to run the length of a hedge before coming to the gate. On the ploughed field my feet became bogged down and thick with clay. It was like running in concrete boots. Even at such a long distance, I could see something was wrong. The crow had flipped Lexi on to her back and had a black, curling claw around her head. The crow began calling. From the tops of the far tree line, what looked like a swirling mass of black butterflies enlarged into the sky with a roaring noise. A huge corpus of crows, easily 300 in number, all intent on defence. They made it to the hawk far quicker than I could manage. On the ground, they circled around, taking turns jagging and pestering the sparrowhawk like vultures or hyenas tugging at the body of a fallen calf. When I arrived they refused to move or be scared away and formed a writhing black mass, powerful and extremely loud. The moment I killed the crow a weird ripple flowed through the mob and they lifted as one, a thick black sheet low over my head. The noise peaked to a truly scary level. I lay over the hawk and covered my head with my hands. With eerie speed, the noise reduced and I watched the crows break off in groups, retracting upwards in a vortex and with the smoothness of smoke returning back to the coppice. Since then I have a deep respect for crows – they are intelligent, brave and, on close inspection, rather beautiful, but if Girl continues to chase crows when free, without the help of a falconer, she will undoubtedly be killed. To prevent this, I have to change locations. I begin hunting with her in the wood where she will be released, slowly easing her into her last lessons.

Getting my son changed and ready for school is not difficult. He likes being at school. He has an imaginative grasp of human history and human potential. He is interested and inquisitive in all subjects and gets ready without fuss and with excitement. I recognize these qualities as my own. Until I moved to the cottage, I had never left the educational system. School, college, university, Winchester School of Art for an MA, then Cambridge, finally becoming a teacher. I am proud he loves school.

I help him get his vest and pants on, then accidentally put his trousers on back to front. Once they have been readjusted, we get his shirt, jumper and socks on. He instructs me on what we need in his bag and where it should go. Running a little behind time, we get in the car and drive to school. Halfway there I tell him I have forgotten his packed lunch. He scolds me with a giggle and I try to wriggle out of it, blaming the dog, blaming the weather, blaming the pixies.

I lose the argument and we turn the car and head home, arriving at the school late. My son is worried and begins to panic. He runs through the gate without his bag or his lunch. I call him back and he sprints across the playground, his oversized bag swinging and bumping around his legs as he tries desperately to wave goodbye.

Throughout my childhood I had various tricks played on me before school. Once, as I was going to remove English homework from my bag, I found that my books had been swapped for a turnip and a flip-flop. I was humiliated, and the homework was significant and I got into trouble, my explanation met with grim looks and not laughter. My shoelaces were once tied in knots to a chair before school. Being late, I had to cut them free with scissors and go to school without laces. On another occasion I had slept in and was once again late. I was told to hurry up and get going. I cycled the two miles like an Olympic athlete, only to find that it was a Saturday. These stories became family legends, stories told to make people laugh. I tell my son about them, trying to make him laugh; he doesn't find it funny. Watching the way my son panics about school, these stories begin to take on a new meaning and I see them for what they were. I cannot imagine playing these sort of tricks on him. I am not sure why I would want to and, if I did, what sort of lesson he would learn.

Boy's last kill is delivered in such a way that it shows me he has learnt enough, that he is ready to go. Scuffing my feet through a patch of errant potato plants, 'moochers' in a field of oil seed rape, the folded green leaves form an oasis for insects and those further along the food chain. As we brush past, a bird breaks from below the canopy and flies fast in to the sky. Boy snatches the air, his quickness to kill caught in a strange, still, suspended magic. He checks and swings back around. I smile and watch him float about for five or ten seconds. I lift my arm in silent salute. Boy passes me, then turns full circle and lands on my glove with his prize. This has never happened with any other hawk. I dare not move. I cannot move. A trusted perch must never move. He begins to pluck and plume, and soon the little bird is bald. I close my glove around its bandy rubber neck and secure Boy to the glove. This is number ten. The magic number. He is now so fit and calm he chooses, and gives consideration to, the safety of where he lands. I am honoured. In a field surrounded by trees with protective foliage he has selected me as *the* place to be. I sit down in the dirt and help him tear up his final kill, feeding him small pieces by hand.

Boy is easily the best hawk I have ever flown, the personification of pure falconry. With a bond like this, my thoughts move into dangerous territory and I fool myself into thinking I should keep him. I am not bothered by laws: I could hide him, keep him as a secret hawk. I will go on the run, be a hawking outlaw and hide myself in the woods for another month in his company. My mind tacks back, taking a more sensible direction. I am aware of the responsibility I owe him. My last act of kindness must be his release and not continued containment. I swing back once more, trying to find a justifiable way around my desires over his freedom. I know Boy would endlessly keep giving moments of intimacy without asking anything in return. Such generosity is heavily tempting. I try to convince myself he is too friendly, too domesticated to be released, my mind spinning in a devilish small-scale version of the modern world. It is easy to take from nature, a lot harder to give back. When we get home, with a permanent marker I draw three simple words all over my hands and arms: *Let Him Go.*

On her penultimate kill I place Girl in a tree near a stream. It flows 300 yards towards a lake with two small islands. I slip over a fence and run back along the field, hop back over and begin beating the cover towards her. A pair of teal rise, moving fast in the wrong direction. The soft, black-grey and brown flash of a rabbit enters a log pile to my right. The usual small birds break randomly and skim back behind me. A moorhen takes flight, its legs hanging below its body in an odd, rickety style of flight. Near optimum weight for release, Girl is less than enthusiastic and watches it for the first forty or fifty yards. The moorhen slows, plops down and walks, its little head bobbing. This change in movement, its complacency, stimulates the hawk, and Girl skims low over the floor in fast pursuit. The moorhen tucks into the safety of the weeds and Girl curls up in a tight arch on to a branch above. She shuffles up and down straining her neck over into the cover looking to kill. I run to re-flush the moorhen and it skitters across the water. Catching up, Girl's downward draught leaves circular ripples on the surface as she clips the moorhen and carries it across the lake to the centre of an island.

On a lake similar to this one, my first sparrowhawk, Daisy, caught a coot over water before tumbling in. The coot lashed out and paddled down, pulling Daisy below the surface. She let go but became tangled in weed, her head just visible above the water. I tried to wade out and fell through a foot of water, right up to my chest in thick silt. A sucking, swirling, black cloud of sub-surface flotsam rolled out in thick clouds and putrid, gassy bubbles popped all around. I grabbed a handful of reeds, pulled myself back to shore, ran to an old boat and pushed off. Scooping up the hawk, I dropped her into the bottom of the boat. By the time we reached home, in a weakened, wet state, she slipped into an epileptic shock. In a darkened room I threaded a thin plastic pipe past her tongue, into her throat and down to her crop. Filling a syringe with liquid glucose, I pumped it into her digestive system. Fifteen minutes later she stood strong and erect. I gave her some fresh meat soaked in glucose and a few drips of Red Bull. The combination acted like a tablespoon of peanut butter or a banana for a long-distance runner. An hour later she ruffled her feathers and let out a scream as if to say, *I am not doing that again.*

Fearful the moorhen may have pulled Girl into the lake, I walk into the water, the sediment in this lake thankfully thicker and firmer. The dogs follow, swim beside me, chuffing and coughing in circles, carrying bits of broken stick and lily pads in their mouths. As I reach Girl's island I climb up the bank and realize it is thick with metre lengths of curling bramble, nettle, whippy saplings and blackthorn. With bramble this dense it is easier to crawl. I push through the brown, crunchy thorns, losing my hat on a recalcitrant branch. Through dark shadows I see Girl's chalky pale plumage. She is glowing. The moorhen is dead. The flesh and skin are tough and fatty and Girl struggles to create an opening. I reach in and, using my knife, cut a small hole in the chest. Girl breaks in easily and I let her feed for a few minutes. Girl's desire to carry, her ability to avoid water, to feed on dry land, are assured. Clever Girl! Once she has half a crop, I slide her back towards me on the carcass, twist on my back and sit up with her in my lap. She keeps feeding, oblivious and unconcerned about the difficulties she has caused. I wade back through the water, Girl held high on my fist.

Several hours ago Girl ate a small bird at the very top of a huge lime tree. When finished, she considered me for several seconds, almost goading, then turned and fell into the air, loaded her wings, built up sweeping speed and disappeared. The wind, high overhead, and the gilded golden clouds of a nearing sunset kept scudding, kept going and going in the same direction as Girl.

She takes me back and forth across the fields, refusing to return. In total frustration I launch clods of earth at the dogs and, cursing the world with increasingly wicked words, continue to follow her across what feels like the whole county. When I get close she takes off and flies round and round, height and the sky the enemy. Now at dusk, and at the end of another two-mile run; bile and breath fight for space in my throat. I am tiring; I have the empty, dizzy feeling that comes with lack of food. She moves a final time, to the bottom of a steep hill in thick woodland. I have climbed up and down this hill three times today. Dutifully, I follow and stand no more than fifty feet below her position. Once again she refuses the lure. On the edge of darkness, I check the signal from the telemetry, note her silhouette and turn tail. She can bloody well stay out overnight for all I care!

Despite my restless hatred towards Girl's behaviour and cramp coiling through my legs before sleep, the drive to reclaim her before dawn is fast and far-reaching. I rise fully clothed, do not wash or brush my teeth; there is no coffee. At 3.30 a.m. I am back on the land in total darkness, a fox-like falconer moving silently through mist-grey, worm-sodden steam and shifting shadows. I arrive at her roosting site, step over fallen trees, drop down into the valley, disturbing strange thermocline bars of warm and cold air. I flip a switch and the dawn chorus I create is a single synthetic beep from the telemetry. She is still here, I just cannot see her. So I wait.

A canopy of trees before first light is compact, a smooth sheet of black in front of dark blue. As dawn breaks, the entropic qualities of light begin to work. At first the trees are a dense tessellation of leaves, a thin, threading lace doily. Then jutting sticks appear left and right like black cracks on a sheet of ice. Less angular shapes arrive: an acorn, a pine cone, a nest, a pigeon, a hawk! To my left the real dawn chorus starts with a single note from a wren. Girl joins in and calls, says hello and moves. Then moves again. Restless and ready to hunt, she flies around in a squashed square. I follow beneath her bell, tripping, tangled, cutting myself and stumbling. She ladders down different trees and makes several stabbing flights at unseen birds below. I get within fifteen feet and throw out a whole dead quail. The pale, bald pink flesh is exposed, the guts and slivers of liver a black rope over bramble. Girl pauses then makes the reluctant choice to return. Watching her feed, I am very weary and slightly ambivalent. It is time for her to go. She is ready.

Release

A week before release I rest Boy and Girl and feed them the highest human-grade meat available. On a diet of duck and pigeon, they spiral to their highest weight. The extra few ounces will give them a week's grace before they are too weak to hunt. It is critical that they kill and feed in this time. Given their joint performances, it is unlikely they will fail. Leaving them to feed alone, I cease interaction unless it is necessary. Without the twang of appetite and with only brief human contact, by the end of the week both hawks are fidgety, more alert and fearful. They gradually become less enamoured of my presence. Girl was never that keen in the first place and I have no concerns she could ever be too friendly. Watching Boy disappear, reverting back to his natural state, is a relief.

On the final night I bring them indoors and spend a couple of hours looking at them, smelling their feathers, remembering. When darkness arrives the fire flickers, casting dull shadows of birds across the wall. I toast them in a happy, sad celebration. I feed Flash and Etta roast chicken. I take Boy and Girl to their night quarters for the last time.

On the morning of release I move quickly. Casting Girl in an old pair of tights and wrapping masking tape around her shoulders, I cut off her anklets, remove her bell and place her in the back of my Land Rover. Although he is reluctant and stroppy, I carry Boy on my gloved hand while driving the quarter of a mile to the release site.

I have invited a small group of locals to witness the moment of their freedom. Nervous with energy, I turn up early and they have yet to arrive. I place Boy on a portable perch then move to prepare Girl. Opening the back of the truck, I feel a flutter and a line of pale grey passes on the outer edge of my vision. A light breeze rolls the empty tubed tights to the floor like the shed skin of a snake. Ever wild, the Houdini of hawks, Girl has done it her way. Girl sets herself free. The fierce, clever, angry Girl is gone.

When the others arrive they tell me that on the walk up they saw her for a nanosecond, passing down the path, skimming over their heads on into the woods.

I laugh.

'That's my Girl!'

Lifting Boy from the perch, I ask the youngest child of the group to hold him while I cut away his anklets and bell. We all look at him and I give a nod. Boy is set free. I expect him to bolt fast, to strip away quickly into the distance. Instead, he twists up and lands in a tree just above our heads. He remains in the same place for twenty minutes. I wave my arms and make a 'shooing' noise. I tell him to piss off. He doesn't listen. Short of throwing twigs, I have no idea how to make him leave. It is a little bit embarrassing. I stare at him as my guests take photographs and videos. The release of Boy and Girl is not panning out how I imagined. It certainly isn't the ending I envisaged, not the type of ending described in the delightful books about released otters and lions I read as a child. Then again, what did I expect? Soft-focus cinematography with both hawks flying into the sunset as end credits roll? If so, in their anarchic reality, Boy and Girl have not read the script. The collected group do not care and are rightly in awe at being this close to a wild hawk. Thanks to Girl, I have had a lifetime of staring at hawks in trees. Getting bored, I make my excuses, say goodbye and drive home.

Back in the cottage I make a cup of tea and go about my business. The time arrives when I would usually go hunting. It is quiet, I hear no bells, there is an emptiness of sorts. My routine has changed, but that is all. I feel no sentimentality, no loss or the expected sense of sadness. Ultimately, Boy and Girl did not belong to me, they were only ever a temporary loan. I decide they are not lost, just relocated. I have the keys to their world any time I wish to visit. All I have to do is open my door, sit quietly, watch and wait. This knowledge is enough.

I am sitting in the soil digging with my son. I have an hour or so left before I have to go. I feel anxious, distracted. I always do. Maybe he picks up on the energy. Maybe he doesn't. Either way, he looks straight at me and says in a voice almost too quiet to hear:

'I am sorry for making you stay.'
His logic is a razorblade of guilt.
'You're not making me stay, mate, but I do have to go. I'll be back soon.'
It is not enough.
What choice does he have, anyway?

Boy and Girl are eventually seen at different locations in the village. My neighbour knocks at the door. She tells me there is an injured sparrowhawk on the path. I walk down to the stream and take a look. There is no hawk. She describes it as 'a small hawk, like the one you had. Its wings were spread open and it was lying flat on the floor looking at me.' An untouched wild musket will not tolerate a human being that close. There is no doubt in my mind that it was Boy on a kill.

Girl keeps her distance too. I only ever see her ghosting through on evening patrols. At other times I find pigeons on the ground for me to eat.

Soon after the release of Boy and Girl I receive a text:

'It's hatched, thin head, stumpy legs, ugly... looks like ET.'

Roughly thirty-six days before this text arrived, in the early dawn, a perfect, pure white egg was laid in the north of England. Three more followed. Each of the eggs was marked delicately with a pencil by the breeder: A B C D. Mine, third in line to the throne, was labelled with the letter 'C'.

There is only one indigenous hawk powerful and capable enough to consistently supply food for free. The little chick with the thin head, the one that looks like ET, is a male goshawk: the same sex and species as the one flown by Haider in Pakistan. Now that Girl and Boy are free, I will use this goshawk to return to the source of all falconry, to its baseline, its original purpose and point. The only meat I will consume from now until the end of the season will be caught by this hawk.

5

'CC'

Summer

I am sitting at a small wooden table, eating ice cream with my son. He has strawberry. I have mint choc chip. His mother has gone off to get a cup of tea. I tell him about my egg and the chick that has just hatched. I tell him about Pakistan. He talks about learning about different religions at school, and we get sidetracked by different ways of life, cultures and people.

'Are people born religious?'

'No, we are born animals and pretend to forget.'

'What's a Mohammed?'

'A prophet. A bit like a Jesus, except he flew falcons and didn't go fishing or bake bread.'

'Do you believe in God?'

'Not really. God seems a bit obvious. It's not nearly beautiful *enough to explain it all. I don't really believe in anything religious, to be honest. Do you believe in God?'*

'Not any more, but I believe in heaven and hell.'

The questions keep rolling. His mum returns and we eventually get back to my egg and chick. I ask him to think of a name beginning with 'C'. Without hesitation he bounces back, 'Chief,' and I reply, 'Catcher.' The little chick has a name: Chief Catcher, or CC for short.

This is the first bird of prey my son has named. It will be the first hawk he has seen in flight and the first hawk he has seen hunting.

Imprinting

There are a variety of different ways to raise a hawk. Some are easier and more commonly understood than others. The history of falconry, its literature, novels and a large proportion of the falconry hawks used in the West are parent-reared. These hawks arrive at the falconer's fully formed, having spent the first part of their lives raised by adult parents either in the wild or bred in captivity. The isolated nature of their formative weeks, without any form of human interaction, creates high levels of fear. Locked away through this fear, a hawk's true nature is muffled, skewed and partially hidden. Similarly to Girl, parent-reared hawks possess an 'otherness', a separateness that remains difficult, if not impossible, to bridge. They have no real sense of loyalty or attachment, and very few falconers manage to overcome this barrier. Even the most observant human will only ever understand or relate to a parent-reared hawk through an opaque veil, guesswork and the prism of misplaced anthropomorphism.

CC is very different.

I will raise him using a method that contrasts significantly with parent-rearing. My aim is to imprint him. The process of imprinting a hawk is time-consuming and difficult to get right. As a method it is ancient, arguably first attempted by the *berkutchi* of Kazakhstan. As a process, it did not travel well through history, or along the trade routes where fully grown hawks and falcon were the norm and far easier to transport. Consequently, imprinting is little understood by people outside avian and scientific circles. Even within the small world of British falconry, only a handful of falconers regularly attempt or use imprinting consistently as a method of training.

If my journeys abroad were about merely observing hawks, and the release of Boy and Girl took me closer to a wild hawk's natural context, then CC and the process of imprinting peel back the layers further, immersing me completely in CC's world from the moment of his conception.

In its simplest terms, imprinting means hand-rearing a hawk from a chick through its juvenile stages right up to adulthood. From such a simple concept a whole new world of raptors opens up. In the first few days of CC's life the plasticity of his brain means that whatever he experiences directly will be regarded as normal. His natural fear for these things is short-circuited and eradicated. As he grows, he will imprint on me, the dogs and his localized environment without learnt fear locking away his more subtle emotional range. CC will behave naturally around me from the outset and will view me as a surrogate parent, a surrogate father. He will become a surrogate child. The relationship I create with CC will unleash his purest behavioural mechanisms and their myriad forms of expression. Over the weeks and months of CC's first year our relationship will be one of undiluted reciprocity, one that pierces the veil and blurs the boundaries between hawk and human so that my life and CC's will concertina, creating an almost transcendental bond, one that rises above and beyond servitude.

Such is the power of imprinting, such is intimacy of the relationship between hawk and human, that the ultimate private behaviour of any raptor is accessed directly. As an adult hawk, an imprinted bird will display, begin calling and initiate mating rituals, start nest-building and either donate semen or stand for insemination in the presence of the falconer. CC's parents, Arthur and Vivien, are imprint hawks, and he was conceived using this method.

Between the months of April and July weird science happens in the private homes, bedrooms, garages and backyards of

Britain. In these urban spaces secretive natural behaviours and bloodlines have been studied and manipulated by individuals whose daily jobs bear no relation to what they achieve with birds of prey. Bricklayers, electricians, chefs, IT consultants, the unemployed, the self-employed, historians, artists – men and women of all cultures, all backgrounds – have developed, refined and successfully bred the best and purest of Britain's indigenous hawks.

The captive breeding of a goshawk is, theoretically, straight-forward. Place a male and female together in a large chamber and wait. In April, they will begin displaying and nest-building and copulate, then the female will lay eggs, incubate and hatch them. The chicks are left with the parents and they are raised naturally – are parent-reared.

The reality is far more difficult. When humans act as an inter-mediary in natural processes, propagating new life becomes difficult. The captive breeding of hawks can go wrong at any stage. The male may be killed by the larger female, eggs may be smashed or tossed from nests. The chicks can be left to starve, eaten, suffocated or knocked to the floor. Bacterial infection, illness and infertility are common. If any of these factors occur, the breeder has to wait another year to rectify any mistakes, by which time a different issue may arise and another year is often wasted. It was in an attempt to circumvent these issues that the process of imprinting was developed by falconers in the West.

CC's breeder, Steve, remains my oldest falconry friend. We relate to each other from opposite ends of the emotional scale; we have very different personalities. If a Venn diagram were drawn, then we exist as friends in the overlap marked 'hawks'. Restless and inquisitive, I went travelling to learn more. Steve, a university lecturer with a scientific mind, moved into breeding. He is the most methodical, calm and consistent falconer I know.

Two or three years before CC was laid his parents were separately imprinted, trained, flown and hunted by Steve. When mature, and as the breeding season rolled around, both were placed in large separate aviaries next to one another. When spring reached full flow, powered by day length and temperature, hormones squirted through their bloodstreams, building in intensity, pushing them both into breeding condition. Deeply imprinted, as they transitioned, Steve began his courtship by taking gifts of food, behaving, touching, calling and interacting with Arthur in exactly the same way as if he were a female goshawk. Stimulated by his presence and behaviour, Arthur returned the favour, calling and displaying to Steve, as if he were a particularly odd-shaped but very sexy female goshawk. Over several days, their courting dance complete, Steve presented Arthur with his clean, plastic-gloved hand. Arthur, excited and accepting, hopped on to it, wriggled about until he found the right spot and pressed down, leaving a single shot of goshawk sperm on the plastic surface. Steve used a rubber-tipped pipette to delicately suck up Arthur's semen and deposit it in small amounts into several hollow glass tubes, then kept them clean and protected at a stable temperature in a fridge.

Running concurrently with Arthur's courtship, Steve began impressing Vivien by building the beginnings of a large, suitable nest. In his presence, as she would in the wild, Vivien redesigned and changed it to suit her own particular specifications. As she was doing so, Steve presented her with gifts of food, wooing her in a similar manner as he had Arthur. This time, the opposite happened: Vivien perceived Steve's actions as those of an odd-shaped but particularly handsome male goshawk. Pleased to see him, she fluttered her wings, bowed and began making a loud, mewing, kitten-like come-hither call. When her hormones hit full peak, she reversed up to the edge of the nest, raised her tail, parted her under feathers and exposed her cloaca (vagina) to Steve. Stroking her back, applying a similar pressure as if Arthur were mounting her, Steve pressed the tube of Arthur's semen delicately against the pink flesh of her cloaca. With sexual satisfaction, she sucked up Arthur's semen.

To maximize the chances of fertility, the interaction with Arthur and the insemination of Vivien happened twice a day, once in the pre-dawn, and then again in late evening.

Inside Vivien, the first half of CC started off as a small, permeable polyp moving down his mother's birth canal. The second half of CC, in the form of Arthur's semen, wiggled its way naturally inside the polyp. A soft shell formed around the fertilized inner egg and a complete CC continued to move down, hardening and sealing in preparation to be laid.

In total, Vivien laid four eggs, each bright white and the size of a small hen's egg. Because Vivien had total trust in Steve, she allowed him to swap her eggs for counterfeit dummies, thus removing the possibility of her smashing or damaging them in the nest. CC and the other eggs were safely transferred into an incubator and kept at exactly 37.4°C degrees, a temperature corresponding to the one found under Vivien when she was incubating naturally.

As the days passed, Steve shone a strong light through the shell of each egg in order to track its development. CC appeared as a dark shadow, a blob coagulating and enlarging as he formed. CC's egg was also weighed daily. As he grew, his egg needed to consistently lose fifteen per cent of its moisture every day. Any more or less, and the humidity in the incubator would have to be micro-adjusted to keep him developing properly.

At thirty-four days, and too big for the inside of his egg, CC wriggled and poked a hole through a thin membrane to an air sack inside the egg. This first breath was perhaps the most significant he would take in his life, as the little bubble of air stimulated a small muscle in the back of his neck – a muscle used only once and specifically evolved to help him break free through the solid shell of his egg. If there weren't enough air, CC would be too weak, suffocate and die. This first breath – his first fight for survival – and the act of breaking free took him two days. When he arrived in the world CC was a bedraggled creature about the size of a credit card. When dried off under a heat lamp, he was, understandably, hungry, so Steve fed him his first meal: thin slithers of chicken thigh. This lightweight meal was highly nutritious and contained enough protein to stimulate all the bacteria and enzymes CC would need to digest all the meat he would consume throughout his life. Once fed, CC was placed in a small plastic crib with a red heat lamp suspended above him to keep him warm and dry. This was the point at which I received Steve's text.

Even using artificial insemination and carefully controlling each stage, only two of the four eggs Vivien laid were fertile. Of the two hawks hatched, only CC survived. CC's sister, Red, didn't make it and passed away after a week. CC was the culmination of six years of effort, learning, love and money, and the first full goshawk Steve had bred.

At twenty-one days old, CC entered his first key developmental phase. He wobbled and stood straight for the first time. He did this independently, with style and with strength, proving he was healthy and would survive. At this point, Steve told me to come and collect him. When I arrive, and knowing my plans, in an act of unstinting generosity, Steve gives him to me for free.

—

My son is curled up in a ball in his garden, laughing hysterically. The dogs are barking and running about. Overexcited and playful, Etta straddles his back and Flash mounts his head. In an act of playful dominance rather than out-and-out sex, they are trying to hump him. They see him as a minor character in the pack, are trying to push him down the pecking order. He is having none of it. He starts shouting at Etta through his laughter. She ignores him and continues to arch her back and thrust her hips like Elvis Presley. She looks around at me with an almost confused expression, as if she has no idea why she is doing it but knows deep down she must. Flash, younger and easily distracted, runs up the garden after the cat. Extricating himself from under the dog, my son points at her, scolding and says:

'Etta, stop it. You haven't even got a willy.'

My son shows a remarkable knowledge of biology. He is deeply inquisitive about where and how he came to exist. He knows about the human body. Nothing is off limits. He has many questions about sex. He shows off his knowledge, proudly explaining about eggs and sperm. He knows how life is made and how it will fade. He knows how CC came to exist, and we draw parallels. His mother and I are fairly honest about it all.

When it arrived, the urge to have my son was strong. Up until my mid-thirties, I never really wanted a child, and then, click, something changed. It was odd, like a male version of a biological clock ticking. This surprised me. He was absolutely not an accident, but there was not too much planning either. Even if his conception had been planned with precision, my deep-seated, dormant reaction to his birth could not have been factored in. If I had known how difficult it was going to be, would I still have had him? The question is a moot point. His mother was adamant. There was absolutely no other option. Her body. Her choice.

Wrestling with him in the grass, the dogs once again trying to join in, watching his beautiful face and expressions, seeing what he has become, I am excited about what he will be. His mother was right not to be afraid.

Later, I ask my son if he wants a girlfriend.

'No way.'

'What about a boyfriend?'

'No, thanks.'

'So what will you do when you grow up? Will you get married?'

'No, I'll buy a house and build an apartment for Mummy.'

'What about me? Who is going to help me when I am old?'

He screws up his face. He thinks for a few seconds, wavers between telling me what he actually feels and what he thinks I want to hear. Finally, he tells me the truth:

'You can look after yourself.'

I laugh. Well done. He does not owe me anything. His life is, so far, his own.

Knowing no creatures other than humans, when CC meets me he shows no fear, just a deep enquiry at my movement and presence. He is roughly the size of a grapefruit and is covered in white fluff. I touch it and, far from it being cloud soft, as it appears, it has the texture of soft foam. Dotted across his back, the tips of his juvenile wings and tail spike through with wisps of gossamer-thin tendrils curling on the end of each shaft. Similar to all accipiters (goshawks and sparrow-hawks), at this stage his eyes are a liquid cornflower blue, his legs already as thick as a human finger. When he stands and stretches his stumpy wings, his toes and long, dark talons stretch flat and far wider than his shoulders. Even at twenty-two days old, he is easily twice, perhaps three times, the size of Boy and Girl in maturity. As I stroke him, he calls, twitters and chirrups with a delicate, soft voice. His head rotates in a smooth motion, following the movement of my fingers. Depending on how close, far and how quickly I move, his irises flex from a pin dot to large black discs.

Before I leave, Steve, his wife Hollie and I sit on the front lawn and have a cup of tea and take pictures. Along with CC, Steve has also bred a tiny merlin called Sir Percival. We place CC and Sir P together on a blanket in the sun. Many times smaller than CC, Sir P shows no appreciation of the difference. He launches himself at CC, tugging and tearing at the tips of his emerging feathers. Unable to get sufficient grip, he becomes petulant and angry. Sir P flaps at CC and tries to drop-kick him into the grass. Becoming annoyed, CC stands up and takes two precarious steps over to Hollie and curls up in the comfort of her soft tracksuit bottoms. I continue to wiggle my fingers in front of his beak, and he plays and pecks at my fingertips; the sharp tip and thin razor edging of his beak, already propelled by tough muscle, cut hard and feel sharp against my soft skin. Becoming bored with my interference, CC eventually tucks his head under his wing, preening free, dry shafts of feather casing that float downwind like fish scales falling through water.

On the drive home Flash and Etta peek into CC's box. He reaches up and pecks and pulls at their muzzles. The dogs jump back across the seats, excited, and bark. CC tracks their movement back to the boot. Content they are at a safe distance, he shuffles over to the edge of his box and sleeps. From this moment onwards, every action I take will be for this hawk; every aspect of his life will be carefully considered and monitored. From the moment we arrive home, I am intertwined with this soft machine of bones, beak, feet and killer instinct in a way that I was not with Girl and Boy.

We arrive at the cottage in the late afternoon and I set to work on my first task. Near the cottage there are numerous wild goshawk and sparrowhawk nests, each one an architectural marvel. The materials, positioning, height, direction and location are never less than perfect. The nest I build for CC is serviceable and safe but in no way as evolved or comparable to the home he would have had in the wild. In the bottom of a tub I overlay a spiral of fresh-cut leylandii. Each green, fan-sized branch has qualities that make a perfect, portable man-made nest. A goshawk nest has to have an uneven surface and not be overly comfortable. CC needs to grip and hold the leylandii in order to move; he should not sit in the same place for longer than is healthy. If his nest is too comfortable he will remain in one position and his bones and muscles will squash up like the bound feet of a Japanese concubine.

The foliage must remain fresh, with small gaps between the fronds, be non-toxic and in plentiful supply. As he proved on the journey home, CC is messy. As his adult feathers continue to emerge, I know his flaky, dried casings will split and fall from his body like rough, circular mounds of dandruff. The gaps between the branches of leylandii allow these husks to fall and collect in the bottom of the tub rather than remain in his nest or become unpleasantly mixed in with his food. When CC eats he will be fervent and toss chunks of meat across his nest, over his head and chest and on to the floor. In midsummer, flies and wasps will buzz about, searching for left-over meat, and with wet mutes (faeces) dribbling over the branches, his bedding material needs to be changed regularly, with a ready supply of new branches.

It takes about an hour to construct his nest and, once he is settled into his new home, he shuffles about, turns in a circle and lowers himself down like a broody chicken. He does not sleep and instead remains alert, following the movement of the dogs around the cottage.

A wild goshawk chick will be brought all manner of creatures by its parents. The feathers and fur of up to twenty-two different species of bird and animal have been found in their nests. They have a varied diet, one that supplies a good grounding in vitamins and minerals – precisely what Girl lacked in the time before she met me. To avoid the issues that afflicted Girl's talons, I provide CC with the food he would have, were I a wild parent goshawk. Unlike for Boy and Girl, who were fully grown, chicken alone will not suffice. So I supplement CC's meat with pigeon, quail and duck. The chicks and quail are easy to access, purchased previously from a repu-table falconry food supplier. The pigeon and duck prove to be slightly more complicated, bought from people who shoot wild fowl, but with the proviso that they use steel shot. Any trace of lead shot is spectacularly deadly to all birds of prey. A small shard buried in the flesh of a duck or pigeon and no bigger than a pinhead would quickly kill CC. Taking a fresh dead pigeon from the fridge, I peel the skin back and check for bruising. I pop the thin membrane of the dark maroon chest with the tip of a knife and poke about in each hole, pinging out the steel shot. Even if these birds have been killed using the correct shot, they may have previously been winged by a hunter using lead. I cut the breast flesh of each bird into smaller pieces, checking and re-checking before washing the meat under a tap, just to be sure. Until his feathers are fully grown, CC needs a constant supply of food: twenty-four hours a day, seven days a week. And all of it has to be prepared fresh and to the same exacting standards.

Eighteen to twenty hours after eating, all hawks produce a small pellet of undigestible matter (a casting) in the form of feather or fur. The casting is a clever evolutionary device that keeps the gullet clean by stripping excess fat from inside the stomach and throat. A casting will also benefit me in my approach to raising CC. Like the Pakistani falconers examining the mutes of their hawks before hunting, examining a cast pellet will help ascertain CC's health. If he produces a tight and compact cast, he is healthy and digesting food correctly. If he produces a cast that is soft, mushy or discoloured, that smells or contains meat, it indicates possible illness. So into CC's fabulous mixed-meat buffet I add small amounts of bone, chopped feather and a sprinkle of vitamin powder.

If I hand-feed CC until he is a fully grown adult, then, potentially, he will become highly aggressive around food and constantly scream. In the wild, CC's protective aggression and noise is the perfect mechanism to ensure that he will be fed by his parents over and above his siblings. At a later date, as he reaches adulthood, this aggression around food will also force a parent to cease feeding and push the youngster from the nest site. Such aggression is a deeply ingrained, primitive part of any goshawk's psychology and it is almost impossible to eradicate completely. Nonetheless, a fully grown, hungry, protective two-pound goshawk in possession of steel-hard three-inch talons is a dangerous creature. Without the luxury of parental rejection, I will have to work through this aggression when it arrives. Until then, instead of hand-feeding and creating an unworkable food association, my only option is to place his food bowl next to him while he sleeps and to interact with him only after he has eaten.

I watch as CC's little white body lifts and drops as he breathes. He twitches and wiggles his head as if dreaming. Maybe because of the scent, maybe because of the noise I make, he wakes, scopes out the nest, notes the change in his surroundings and begins twittering with excitement. He gets uncertainly to his feet, wobbles and runs forward, falling face first into his bowl. Spreading little stumpy wings like a bat crawling up a cave wall, he twists about like an excited child, scrambles back up on to his feet and plunges his beak violently into the bowl. He has a voracious appetite and a ticking, beeping call emits deep from within his chest as he picks and jerks back chunks of meat with a passionate gulping. Some pieces of meat are too big and hang over the sides of his beak, so he flicks his head, drops it to the floor and tries again. His crop eventually fills to become the size of a small satsuma and his jerking, tense movements pass into a slow, sated, soft, delirious contentment. His tight, defined outline relaxes and he blurs to a rotund bubble of a hawk, shuffling about his nest. I reach in and pick him up and sit with him on my lap. I stroke him like a cat, and the dogs come over to investigate the meat-scented ball of fluff. Finding nothing of value, they head to CC's vacated nest in search of left-overs. CC eventually ceases shuffling, fires a mute with accuracy down my leg and on to the floor then settles to sleep.

To immerse CC totally in the world of humans, I carry him everywhere I go in a portable nest (a large basket 'borrowed' from a supermarket with leylandii branches in it). His wide, flat head with its blue eyes poke over the edge at the strange new worlds he encounters as we take to travelling. He comes with me to the bank to pay a council-tax bill, his charm quickly offset by the spurting white stripe he leaves across the carpet. In the local town park I lift him out of his nest and let him walk about in the grass. He makes for an unusual spectacle. We are quickly surrounded by an inquisitive group of adults and their children. Through CC, I find it easy to relate to the group, with their rapid-fire questions. I feel relaxed discussing his origins and future, the parameters of interaction safely set by the fluffy life form in front of us.

On one or two warm late-summer evenings I walk through the fields with the dogs, CC tucked in a hawk papoose (a box tied with string around my neck) to a pub by the river. It is a popular location. The same reaction occurs: people pick him up, stroke and touch him. All ask questions. CC displays no form of distress, only a laid-back, placid acceptance of each situation. His mind is sucking up information, normalizing the human world, almost as if he were human himself. Quickly bored with the company, he waddles off under a wooden table and falls over. A group of bantam chickens see the shape of a hawk and scatter away down the bank to the river. Swifts and swallows come screaming in through the garden, check CC out and wheel up in a circling mass above our heads. A misguided lady leans down and attempts to feed him a chip; he grabs it from her hand and flicks it to the floor. I walk over and save him from eating his first (and only) vegetable and gently guide him away from further trouble.

A small goshawk chick will make an easy meal for a buzzard or sparrowhawk and cannot be left unattended on a lawn. At the cottage, as the days pass and as he grows, I place CC in a large square pen outside the front door. I watch him inside his pen while I paint, allowing him safe access to sunshine and fresh air. He readily hops out of his nest and stumbles across the AstroTurf exploring the extra space. As he develops strength, he begins branching out further from his nest, leans forward, stretches his wings and trips over his feet. As the days pass, each time he falls, he rebalances and stands up more and more confidently building up to the next key phase of his existence. After a week of continued collapsing, stumbles and learning, I observe a profoundly different type of movement.

The one aspect of a hawk that fascinates me over and above all their other physical characteristics is their feathers. More so than hair, nails, scales and fur, a feather is the single most complex organism growing from the skin of any creature. Feathers have evolved not just for flight but to help in a multitude of other behaviours. They are plumes for camouflage, defence and sex. Feathers are to keep a bird cool or warm, confound or confuse a predator, make or muffle noise, improve hearing, line nests, carry water and ease digestion. Looked at under a magnifying glass, CC's half-grown feathers are no less impressive. Designed for predation, they are intricate, indescribably complex and astoundingly beautiful. They grow in tight, straight lines, have whorls, tubes, barbs, flat, parallel strands, hooklets and asymmetrical veins. It is with these astonishingly evolved feathers that I watch CC, within his pen, rise above gravity for the first time. He jumps and lifts off the ground, half hovers then descends back to earth. His first flight may have been brief but it is met with elation by both of us. I smile and CC takes it upon himself to run and flap the full length of his pen, as if he has just scored in the World Cup.

With all this growth, energy, exercise and a total lack of fear, CC becomes highly mobile and his explorations expand more and more. I have to run some errands in town, so I tuck him up in his nest, leaving him on his chair next to Flash and Etta. When I arrive home, the dogs are still asleep in the same position but CC is nowhere to be seen. I can hear him twittering, and Etta lifts her head and sighs. I look over her shoulder and find him nestled behind both dogs on the sofa. Finding the fur of the dogs more comfortable and certainly a lot warmer than leylandii, he has decided to relocate himself.

I pick him up and put him back in his nest. He throws a violent tantrum, screams and wriggles in protest. Twenty seconds later, he launches up out over the top of his nest, makes a strategic bounce on the arm of the chair, a flying flap, then a soft thud back behind the dogs. It is one of the funniest things I have ever seen a hawk do.

Scared he may be accidentally crushed, I force a compromise and give him his own cushion on the arm of the sofa, near but not on the dogs. At first, he takes to it with alacrity and remains in position. He then starts to push the boundaries, branching back out into the cottage. There is no compromise or warning, and he suddenly lifts up a few feet before dancing up and down on his cushion. Flapping harder, he rises higher and higher, then lands hard before running the full length of the sofa, hopping across and clawing the backs of the dogs then boomeranging back and footing his cushion as if it were a kill. The dogs huff and puff and move upstairs for respite.

When he has finished exercising, not wanting to sit, CC takes to climbing and flapping up a chair near the back window of the cottage. He spends a lot of his day contentedly looking through the glass, studying the wheat moving in the wind. He twists and turns his head almost upside down, a specific developmental movement of young hawks that builds the muscles of his iris, calibrating his pin-point eyesight by using the tops of the crops as a target.

Whenever I sit down he immediately hops on to my lap and begins to nibble and play with the cloth of my shirt. He takes to using his extraordinarily long talons to climb and crawl his way up on to my shoulder. This is the best position to begin chewing my hair and pulling at it, as if I am the corpse of a bird.

At night I take him upstairs and place him next to my bed. As the dawn arrives I hear him rustling about in his nest as the new day forces him into action and he becomes restless. On more than one occasion he wakes me by landing on the pillow and shuffling up next to my face. A burning sensation cuts across my lip or cheek as he rudely wakes me with a very sharp nip on the face.

It is misguided to interpret this running, jumping, flapping, flicking, fighting and footing as simply the playful exuberance of a young hawk. His behaviour is quirky and charming and would be easy to anthropomorphize. None of it is for fun. All his actions are part of a deadly serious, evolved preparation of mind and muscle, for the battles he will face in adulthood.

—

I am alone in the garden, watching my son play. His capacity for imaginative performance is extensive. In each hand he has two spindly 'T'-shaped pieces of plastic. He has constructed them using pieces from a Lego-like toy called K'nex. They are minimal; if I try hard, I can just see a human-like form, two arms and a central body. He balances along the brick edging of the driveway, flying each figure through the sky. He makes huge whooshing noises and I can hear small snippets of conversation. 'Nooooo… that's right… fight them.' A battle, perhaps. There are more explosions as he heads up the garden and disappears around the corner. A few minutes later he returns, tiptoeing along the bricks and making odd bleeps, bangs, machine-gun noises and a wide range of other sound effects. He is so interesting I find it difficult to ignore him when he is in full flow. He spots me watching and becomes annoyed, self-conscious. Tells me angrily to stop looking at him and moves off again, away from prying eyes, back up the garden, blasting imaginary enemies out of the sky.

When I was still in primary school I was left alone for large portions of the day throughout the long summer holidays. I loved it. This was when I was allowed the freedom to explore the countryside, but I also, for some odd reason, found cookery books fascinating. I have a memory of a thick book with a smiling woman on the front cover. I would pick a recipe and follow Delia Smith's suggestions, baking and cooking different foods throughout the day. I also had a predilection for making my own imaginative creative operas and performances. I would endlessly play records from my parents' vinyl collection, performing various scenes to the songs by running between different chairs wrapped in towels of different colours. I would draw and paint, read comic books and create weird and wonderful fantasy worlds reminiscent of the films of Ray Harryhausen.

When my son was about three, I remember taking him to a junior playgroup. There were four of us: his nanny, my father, my son and me. The pressure to perform was immense. Midway through, the supervisor started telling me how to interact with my son to maximize his 'developmental and social skills'. The nanny was sitting next to me and my father was poking a camera into our space, taking pictures. I reached in at the same time as the nanny. It was embarrassing. 'You take him'; 'No, you take him'; 'No, you take him'... Click, click, click with the fucking camera. I hated the intrusion, was deeply self-conscious. I would have been happy just playing with him alone.

My son comes back round the corner and I ask him if he remembers the playgroup.

'No.'

I watch him wander off, destroying some invisible planet on the other side of an imaginary universe as he goes. A universe that only my son knows exists.

—

As CC's size increases, so does his appetite, almost tripling overnight. I find myself refilling his bowl three or four times a day. This is highly unusual and leaves me confused. After preparing yet another meal, I place the bowl in his nest and watch what happens through a crack in the kitchen door. CC ignores the bowl and Flash makes his move. Inching across the sofa, he begins nosing through the nest and eats all CC's diced meat. CC stands no more than half an inch from his face, twists his head upside down and watches Flash intently. Thankfully, CC is not hungry on this occasion. If Flash attempts to do this when CC is on a kill out in the field, the situation will be very different.

The cottage slowly becomes a mess of hawk mutes, fledgling fluff and feather dust. CC is slowly dominating the space we live in. I keep losing him and finding him in different places around the cottage: in the kitchen, tearing at a tea towel; upstairs in an open wardrobe; or chewing twigs, covered in soot, behind the log burner. When I take him outside so that I can clean, he screams repeatedly to be let back in. I have ten minutes to sweep and mop before he fires up out of his pen and re-enters the cottage, strutting and flapping, his ticking talons clicking across the wooden floor.

This level of mobility means it is time to attach his equipment. Cutting two anklets and two jesses, I stand CC on the table with the bits of leather next to him. As I turn around to find the swivel, he leans over and down, picks up a piece of leather and swallows the whole jess. When I look back, all I can see is the knot hanging out of the side of his beak. I grab it and pull it out of his throat like a sword swallower removing a blade. He reacts to the removal of his 'food' with intense annoyance and begins screaming at me, flaps to the floor and rolls around in protest. I eventually attach all his equipment and take him out to a secure newly built caged pen in the garden. Inside is a curved bow perch. I tie his leash to the ring and he clambers up on to the rubberized surface and stands happily. Just in case his legs tire or he wants a nap, I place his nest next to the perch and leave him to it. Unable to see or hear me, he settles into this new routine without serious protest.

With the summer season in full flow, I am asked to cut the lawns for a country estate. Driving to the estate takes ten minutes and on the journey CC perches happily on the back of the passenger seat, staring out of the window. When cutting the long driveway or the cricket pitch, I place his perch in the middle of the lawns and zoom around him with the mower. When not motionless and transfixed by the movement, he lies with his wings and tail fully spread out, sunbathing next to his perch as I whizz past. After a couple of hours we head home, passing crows, the odd pheasant, small birds and rabbits. His gaze locks on to them instantly, feathers tightening on instinct. He is close to adulthood. He needs no lessons on what to kill, his meaning and purpose is in the movement of fur and feather already out there in the fields.

On the hottest day since records began, my ankles are sweating even when I'm barefoot. CC is in shade, his beak slightly open, wings sprawled out, legs akimbo in the heat. He is clearly hot. I pick him up and walk to where Boy had his first wild bath. I step off the path, through the nettles and down to the little stream that arches and curves through the wood. Jumping down on to the gravel, I tie CC's leash to my foot and open my glove. Looking at the silver bubbles, dappled shadows and leaves turning in the current, he comically clicks his beak together; if he had lips, he would be smacking them. He bobs his head, drops on to the gravel, flaps his wings and dips his bottom and tail on to the sand and stones a long way from the water. In his innocent eagerness, he is not experienced enough at bathing to realize he has to walk into the water. I put a hand on his shoulder and shove him into the current. He pauses, thinks about whether he likes it or not, decides he does and resumes his remarkable behaviour, spending the next few minutes rolling and splashing about like a toddler in a lido.

I am soaked in water and my son is in hysterics in a deep bath of water. I am sitting on the toilet, pretending to be a jazz musician, doing a speech to an imaginary crowd. I begin tapping out a little rhythm with some toys: a plastic tube and a small metal bus. Then I point to him, indicating that it's his turn to make his stunning Charlie Parker-style solo. Instead, he makes the worst noise possible and rolls about laughing, splashing water all over the floor. It is a ridiculous game, but he loves it and I love it. It has become a routine, something we have developed together.

Before I took my son home from the hospital we had to be shown how to wash him. This confused me. Coupled with my desire to leave the ward, instead of interpreting it as help I felt that it was being implied that I was in some way incapable, as if to wash a child would not be instinctual, that I would accidentally drown him or just not bother. I wondered if the demonstration was designed specifically for us, for me. I felt infantilized and the words and instructions fell with a sense of condescension. There was a tacit inference, I felt, that 'He's a man, he is bound to get it wrong.' The situation was made even more confusing because I knew the nurse giving the demonstration. I had taught her children. She had five children of varying ages. They were a dedicated, loving family – good people. Yet, without exception, all came to school unwashed and with unclean, stained clothes. The memory of this remains vivid. For some reason, the difference between her private life and her public service position seemed significant at the time.

The discordant jazz mash-up my son and I have created is now over. We can hear his mother's car crunching on the gravel drive and he is excited by her arrival. I have put far too much shampoo in his hair and it takes at least ten jugs of water to rinse it clear. I hold my hand on his forehead. I remember getting soap in my eyes when I was a child, and it was horrible. Thankfully, he is spared this ordeal due to my ability to remember. Making sure he doesn't slip on the floor, I lift him into the bedroom, whizz the towel around his naked body, slide him into his pyjamas and send him downstairs to his mother.

Simple, really.

Near fully grown and unlike the sparrowhawks, CC is too big for the house, his feathers long enough to be broken on the sofa or the legs of the chairs and tables. I relocate him to his night quarters for the first time. In the morning I have to visit the vet so I change out of my normal clothes and put on a new bright blue checked shirt. When I open the door to collect him, he rears up, his eyes wide and aggressive, and he begins to bate away from me furiously. It takes a few moments to figure out what the problem is. Then it hits me: I have changed colour. He does not recognize me, I am a bright blue stranger walking into his space, trying to pick him up. I leave him in the mews and change back into my work clothes. Only then can I calmly transfer him to his perch in the garden. I text Steve and ask him about this behaviour. His reply is straightforward:

'Imprint goshawks are autistic, mate.'

When CC's feathers are just below full size, he is on the cusp of what falconers classify as being 'hard penned'. He is no longer 'in blood'. This means the oxygenated blood carrying hormonal messages for growth recedes from the feather shafts and returns to his body, flesh and muscles. Like the human voice breaking in adolescence, or the obstreperous hormonal arguments of a teenager, CC undergoes a similar change. He becomes agitated, angry and restless. This change would be met with aggressive resistance by wild parents and he would be forced from the nest site. Instead of pushing him away, I have to put up with it and begin to prepare for his training.

When he is fifty-six-days old, I pick CC up as an adult hawk for the first time. The difference between the previous twenty-four hours and this moment is obvious. He feels different: a condensed mass on my glove, he has swift spirit and the whiff of the deadly. The feeling flows down my body, sinking my feet into the ground. It is so very, very clear why goshawks have stood the test of time. He has the fidgety presence of a middle-weight boxer before stepping into the ring. He is in immaculate condition, is roughly the same size and shape as Haider's goshawk and exudes the same static power. His eyes are daffodil yellow, the back of his neck is covered in an unusual deep orange buff, like a setting sun permanently cast across his shoulders. Thin, lozenge-shaped feathers snake down his chest. At the sides, they swell and morph into little brown love hearts across a pale, light cream covering. He has his father's build for speed. His beauty and the crackling potential for unrestrained violence belong to his mother's genes. Relaxed, CC's grip is soft on the glove, but he cocks his head, turns a beady eye, leans forward and lets out a bellowing reptilian scream.

Over-confident, misreading and underestimating his behaviour, later in the day I pick him up without a glove. Unfamiliar with the perch, he clamps down hard. His talons penetrate my flesh easily, forcing themselves into my hand with no more effort than it takes to sink an ice pick into putty. For several seconds my thumb and forefinger are pinned together. A small amount of blood wells up and flows through the folds in the palms of my hands. A boundary has been broken and this first blood flows to a new type of bond. A bond that hurts.

The bond between my son and his mother is a never-ending source of wonder. I have been looking after him all day. To pass the time I took him to buy an art-based board game. There are various objects, films, books and personalities to be either drawn, modelled or sculpted from paper in accordance with the instructions on a clue card. The other person has to guess what has been created before the sand in the egg timer runs out. It was set up on the table hours ago. He has been waiting for his mum to return so we can play the game. She left at about 6 a.m. and arrives home late, around 8 p.m. As soon as she walks in she sits down, the dice are rolled, the counters move and we begin to play.

I show her the clue on the card in my hand without my son seeing. She makes a weird blob out of Plasticine. By any stretch of the imagination, it does not resemble anything remotely like the subject. He guesses and gets it right first time. Again, her turn arrives, and the same thing happens. And again. When it is her turn to guess, my son doodles and shapes a small bit of half-torn, crunched-up paper. I am holding the card with the clue so his mother cannot see it. I struggle to see even the slightest hint that what he has created corresponds to the clue. She gets it right. Every time. It is disconcerting. I tell them they are cheating. They laugh and tell me I am just a sore loser. It takes every ounce of effort, drawing on all my skills as an artist, to try to win. They are like twins. It is a tour de force of mutual understanding, their intuition almost telepathic.

My own mother would have been a remarkable woman had it not been for the men in her life. She remains overshadowed, an emotional shape-shifter, changeable in the face of dominance. I love her, but our bond is ethereal. I find it difficult to latch on, to find a consistent definition. She is there, then she fades to grey. In a certain sense, all my human relationships have been a failed search for attachment. But if you don't know what something looks like, how can you recognize it and deal with it appropriately?

By the time we finish playing two games and pack the board away, it is at least 10 p.m. and time for my son to go to bed. His mother thanks him. 'That was just what I needed after a horrid day.' It was a lot of fun.

When we are all in our respective beds, I think back to when my son was a baby. I begin to compare my feelings now to my feelings then. I am hit by a wall of utter shame. Along with all my other muddled emotions, tangled up and deep down I think I may have been jealous of my son. Jealous of his attachment to his mother, and a mother's attachment to her son. I am thankful that this feeling has passed, transmuted into something far more stable, far more positive. I see the connection between them for what it is. See its value. I am glad my son has such a strong attachment to his mother. It makes him whole. It will hopefully enable him to form balanced relationships when he is older. I think for my part, I have hit too many fences, I do not have the energy or capacity to succeed. I have all I need with the natural world, with my son, my dogs and the hawks that come my way.

Raising CC to his adult juvenile stage is easy. I understand him, can read his moods and know what he needs. Handling his reactions as we move towards hunting condition is another matter entirely. In comparison to a parent-reared hawk, he will need only a small amount of motivation to return. What worries me is that, when training an imprinted hawk, there is a distinct transference of power. The fear CC lacks has been redirected. I am the one who fears for the unexpected, for injury and for my physical being. The charming behaviour he has displayed over the last fifty-five days will disappear and not return for many months. When training starts in earnest, he will have a direct, dominant way of communicating displeasure. He will use extreme physicality: biting, footing, swiping, wing-whipping, flapping and seriously attacking me. There will be no discussion or warning. If my approach is wrong, I will be instantly chastised with force, and in these situations it is difficult to discern who is teaching whom. It is these reactions that make most falconers fearful of imprints, but this aggression is only a small part of the way CC will eventually communicate. His aggression will taper off the more he succeeds when hunting in the field. Until such time, I will just have to work through it, will have to deal with this particular phase of his learning until he begins to respect my company.

For the last twenty minutes my son and I have been having a discussion. It is now bordering on an argument.

'You are basically too young.'

'Well, I have seen an eighteen film before.'

'That was an accident.'

'Well, it wasn't the second time you let me watch it.'

'This is true, but that was also an accident.'

'How can you accidentally watch a film? And it had swear words in it. Adults are supposed to be responsible.'

'I know, don't mention that to anyone. I didn't know it was an eighteen. The bottom line is you're too young, so Grand Theft Auto is out of the question. It's too violent.'

'You've played it.'

'Yes, but I'm forty-three.'

'I'll download it.'

'How will you pay for it?'

'I've got money.'

The discussion keeps on. He is like mercury. His ability to parry and joust my points is remarkable. I am caught between frustration and admiration. He intellectually and verbally punches well above his weight. The truth is, I really do not care one way or the other. He is perfectly capable of distinguishing between what is real and what is fake. Perfectly capable of dealing with the game's content. I have every faith he will not pick up a baseball bat and beat a pensioner to death. The conversation has gently slid from logic and truth into a matter of principle. I notice myself regurgitating words I have heard in the past. It's an odd feeling. I begin to fall into cliché. Begin to assert unjustifiable dominance. I catch myself saying things my father says, what every father says. I have finally turned into a universal father. It makes me laugh, but I also feel a slight twist of annoyance.

Fully aware of my words and of my total abnegation of any type of individual responsibility, I simply say:

'Look, ask your mother.'
'I already did. She won't let me have it.'
He has been playing me as a soft touch all along.

I begin reducing CC's food intake to one meal a day. In the build-up to his daily meal I walk him across the fields with the dogs; midway, he produces a large, tight casting. By the time we arrive home his screaming is so loud it bounces off the trees and walls of the cottage. He turns towards me and screams repeatedly into my face. His feet pulse down and grip so hard they squash the bones in my hand. Despite my thick leather glove, his central talon presses into the ball of my thumb with the same feeling as the blunt tip of a ballpoint pen pushed into a hand. He puffs up and starts to bully me. I weigh him. He grabs the scales and lifts them off the bench, then strikes out at the glove with his free foot. It takes five or ten minutes to get a consistent reading. He is almost at his top weight, nearly two pounds. I leave him another hour, letting the sun drop further down the sky. At dusk, I weigh him once more: he is the same weight, only now raging, screaming, faster, louder, longer, mantling low on the glove. He is telling me quite clearly that he is starving.

He is lying.

Normally, a hawk is called to the glove. With an angry two-pound imprint goshawk, this would be problematic. He may come off the glove and attempt to strike me, so I prepare a lure for his recall. Girl required the lure because she was frightened of me. I intend to use the lure because I am frightened of CC.

I attach a creance to his swivel, place him on a perch, walk away, turn and quickly throw the lure to the floor away from me. CC reacts instantly and smashes it into the ground. Replicating his behaviour on the glove, he spreads his wings defensively, mantling the lure, screaming, trying to threaten me and push me away. The dogs get too close and he moves to attack them. They scuttle away. He rips into his food, screaming between mouthfuls. Crawling on my stomach, low and unthreatening, I reach in with my glove. This provokes no further anger and CC keeps feeding. I slip my bare right hand under the glove and delicately began stroking his feet with the tips of my fingers while he feeds. He looks up, pauses then returns to his meal. Ever so slowly, I remove my glove and reach in with both bare hands and gingerly begin breaking up his meal, feeding him with my fingers. I want to show that if I help him, he can feed easier. That I am not stealing his food, that he does not have to defensively protect his food. As I fiddle with the slippery meat, he stops, steps off the lure and on to my wrist. I freeze. CC's feet tighten, the tips of his talons make a light indentation in the middle mass of blue veins. I suck in, hold my breath, feel the beat of my heart, expecting the searing pain and a trip to hospital. A relaxed ripple of feathers spreads over his body, then they tighten flat. He considers me for several seconds. In deferent submission, I turn my gaze to the floor. With unexpected softness, he steps off my arm and back on to the lure. I continue to tear up his meal. When he has finished he looks about, checking to see if he has missed anything. I back up and sit on my knees. Holding a chick leg in my gloved hand, I offer it to CC, who hops up off the lure and on to my glove. I am shaking with relief and excitement. It is an exhilarating first lesson and one that I've escaped without injury.

Within a few days he is flying 200 yards and hitting the lure hard in a variety of locations. Each time he remains at the place it landed and feeds instantly. In building our routine, my behaviour remains constant and repetitive as I reach in and hand-feed him. The trust builds between us with each successful lesson. Like Boy, he is smart, and progress is quick. Watching him closely, noting his reactions, it is apparent he is ready for free flight.

To fly CC free, he needs telemetry and a tail bell for location. Telemetry is attached using a spring clip. The spring clip is squeezed together and slipped inside a plastic backpack. To attach the backpack, two, thin Teflon ribbons are curled around the shoulders and chest of the hawk, pulled tight then sewn together at the front. It is a fiddly, tricky operation. CC needs to be hooded and another falconer needs to help. It takes Steve, Hollie and I nearly forty minutes to complete the task. We get it wrong, start to argue and bicker. We shout at

each other. Hollie nearly bursts into tears. Likewise, throughout the process CC sits low, silent and angry on the glove. His
wings hang down by his sides like a dead pheasant. As soon as
our joint indignations are exhausted, I place him back on the
perch in the garden and let him take a bath.

In the morning I intend to fly him on the creance for the
last time. When I approach him in the mews, he goes berserk.
On the glove he rears up to full height, opens his wings wide,
clamps down with deadly force and puffs his chest up and
out. The feathers on his head form a broad crest up over his
head and his eyes are bright and wide with alarm. He emits a
long, relentless twittering. It is utterly disheartening. We have
taken several steps back. The backpack fitting has caused a
deep-set resentment which has clearly festered overnight.
The speed of his sensitivity, how quickly CC remembers and
displays his feelings, are remarkable. I understand his anger
and the perceived inconsistency of my behaviour. I have gone
from tender respect to horrendous heavy-handed abuse for
no apparent reason. Steve tells me it would have been worse
if he had not been hooded and seen me do it. The only cure
is time and patience. For three days CC remains resolute in
his distrust. It takes close contact and gentle stroking every
ten minutes before we begin again and achieve free flight, and
start hunting.

Rabbits

It is impossible to explain concepts like patience, summer-time heat, heavy cover or the abstract idea of 'there is always tomorrow' to a hawk. CC's mind and body exist only in the present. We keep missing, or on some days do not see rabbits at all. As we fail I watch him coil tighter and tighter with restless rage, his instincts having no conduit for full expression, every fibre of his being is becoming frustrated. His screaming reaches at least eighty, ninety, a hundred decibels; he sounds like a rusted gate swinging endlessly backwards and forwards in the wind, or galvanized nails scraped across a blackboard. I am convinced that when I am not looking he purposely leans closer to my ear and screams as loudly as possible, telling me to get on with it. After he has been put away in the evenings, my ears continue to ring with a tinnitus-like *eeeeeeee*.

On the morning of the fourth day of failure I weigh CC and he is less than receptive. As usual his one foot clings viciously to the scales, the other is gripping my glove. The scales lift from the table and the leash and jesses become tangled. I open my glove to adjust them and CC lunges onto my belt. With overwhelming speed he drops the scales and attacks me, walking up my abdomen, chest, and shoulders, lunging and footing me as he goes. He stops just short of my face and neck.

When flying and he misses a rabbit, and if I fail to produce the lure quickly enough, he strikes my shoulder or legs with the same level of determination he would have used if he had killed. My bare right hand takes most of the punishment and is deeply slashed and cut by his talons. My middle finger receives a heavy blow, and as infection sets in I am unable to bend it fully without stinging pain. My shoulder and upper left arm are covered in small pinprick dots where he has purposely missed the glove and gone for skin.

In one instance he misses four rabbits in the space of ten minutes. On the final flight, he boomerangs back directly at my head. As he spins out to hit my face, I have to swat him out of the air as if he is a giant angry wasp. Later when his determination to attack is at its utmost and he has once again missed, he sweeps silently behind a hedge, flips over it and smashes into me with full force. There is total silence and I feel nothing. My brain protectively separating my mind and body, I know something is wrong but cannot feel it. I hear a weird squeaking sound, like a balloon being rubbed on wool. His talons sink into the skin at the top of my right shoulder. I turn to face him and I hear a deeper, popping, tearing sound as he sinks right into the muscles. I am hit by a sudden burning sensation. He tightens his grip to full strength and I rotate left then right, trying to escape. Each movement curls into the curve of his talons and his feet go deeper. I have to physically rip him out of the top of my arm removing cloth and a large gouge of flesh and throw him to the floor, followed by his full rations for the day. Feeling light headed, dizzy, and nauseous, I sit down beside him.

The next day the pain is immense, the whole of my right side, from my collar bone across my chest and down to my waist is burnished yellow, green, and purple. It looks like I have been hit by a car.

I text Steve:

"What the fuck is the point of an imprint goshawk?! I am not a sadomasochist. I am in agony."

With oblique, arch knowledge, he replies:

"Use the force, young Jedi. Wait until the freezer is full, then you'll see what the point is. Also, learn to move a bit quicker."

He is right of course. Far from being a problem, this behaviour is to be understood and accepted. It is this innate fearless aggression I wish to harness for hunting. All CC needs is turning outwards, given the chance to function in the world in the correct manner. He needs to fulfil his instincts and celebrate his skills. He needs to kill.

The letter in my hand states:

'The children will be learning about the Stone Ages. We encourage them to come to school dressed as a caveman or cavewoman.'

When we get home we sweep through Google, searching for images. He is certain he wants to be a chief catcher, a tribal hunter. We find pictures of antelope skulls used as headgear and work out a plan. I begin to get overly artistic. Within three hours I have the basic shape of the skull slotted together with cardboard, the horns made of rolled-up newspaper. I drive to a model-making shop and buy Modroc, paint, polystyrene bones, a large roll of fake tiger fur and some plimsolls. I begin the final construction, my son gives a nod or a shake of his head and I adjust my plans in accordance with his requirements. This is our first serious exploration in art and I am in safe territory. I know the intrinsic value of arts and crafts. This is the best of me. I have no hesitation in showing him what I can do. I am driven to impress him, to make him proud of me.

My ability in other domains as a role model is less precise. I do not conform to the general consensus of what it is to be a man. I observe the lives of other men in their forties, see their families and their possessions, and know I am out of synch, anachronistic, an oddity. I know the problems that will inevitably arise as my son lives his life will not be met with the usual advice, guidance or a standard point of view. How he will respond to who I am, to my thoughts and feelings, as he begins to form his own personality into adulthood, causes me a mixture of concern and interest. At best, I can offer a creative, contrary alternative; at worst, I will be totally unhelpful.

At the table I continue to creatively show off. I add Modroc to the mask, shaping it and smoothing it to the texture of bone. I add paint: yellow ochre and white. I brush tones and shadows around the eyes and base of the horns. I go for a walk near the woods behind his mother's house and collect pheasant feathers, twigs and a long length of old binder twine. I cut holes in the polystyrene bones and thread them on to the twine. I add flaps of Modroc to the ends and paint them red, replicating pieces of fresh flesh. I fold the fake tiger fur in half and cut a hole in the centre to form a poncho. With the remainder of the twine I make a belt to tie around his waist. I break the twigs into small lengths and glue them to the plimsolls, making a pair of caveman clogs. For the final effect I cut and shape a spear, tying the remainder of the feathers around the tip. My work complete, I lay it on the table and my son gives a final nod of appreciation. There is no question he will be the best-dressed, hunting, gathering, meat-eating killer caveman the world has ever seen.

We have been out for two hours and seen nothing. I am sweating heavily. Inside my shirt, droplets fall from under my arms and down the side of my rib cage. The hatband on my head is stained black and liquid salt drips through my eyebrows into my eyes and stings. The night before there was heavy rainfall. The corn stubble clicks and ticks, drying in the heat with the sound of a thousand miniature clocks. The dogs are exhausted and whirls of heat vapour lift off the ground. Etta is fussing around my feet, stops and rolls on to her side, panting. I give up, tell her we are off to the stream to cool down.

Up in front, at forty yards, a rabbit materializes out of the corn and runs, sending up puffs of dust down the tractor tracks. I look at CC and wait. He is engrossed elsewhere, cleaning the edges of his wings, and does not see it. Neither do the dogs. I stand straight and raise my arm vertically. I feel a slight bump and strangely weightless, as if I have detached and thrown my arm forward in silence. CC accelerates to fifty or sixty miles an hour, closing down the distance quickly. The rabbit is still running, makes it to seventy, eighty yards. Inexplicably, it stops, turns and looks at the hawk. CC hits it hard and lifts it cleanly off the ground. The rabbit flaps beneath him like a loose sail set free in the wind, and they land way past the point of impact.

Sprinting the distance, skidding on my knees, CC's back right talon is buried deep into the base of the rabbit's skull. It is dead, but still moving. CC has killed it outright. To be certain, I lean in and break the rabbit's neck quickly and cleanly. CC protests at my interference, is screaming and mantling. I move back and let him settle. He pulls hard at the rabbit's fur, stripping light tufts, filaments of which stick to his beak and eyeballs. Frustrated, he rolls his head across his wings, shakes the fur free and sneezes. I lean in with my gloved hand, cut a small slit in the rabbit's chest and pull it apart, exposing the ribs and inner chest cavity. CC is suddenly silent and clamps down hard on the carcass. Once again I step back and let him feed. The blood pools around the rabbit's lungs and heart. CC moves and the rabbit tips. Blood spills over his feet, coats his feathers, drips in small dots and spreads out in rivulets across the earth. I push my hand inside the rabbit's body and feel an intense, wet heat. CC continues to dip and rip into the chest cavity, allowing me to pull apart small pieces of flesh and hand-feed him.

I place a large piece of warm liver in my mouth. It feels like firm jelly, tastes sharp and metallic. CC picks up a blue/grey intestine and begins to eat it. Intestines are not a good meal for a hawk. I grab the end and we have a brief tug of war. I remove the rest of the stomach and throw them behind me, allowing CC to feed on the choicest meat only. The dogs find it and have a tussle. As CC fills himself up, he calms. I reach in a final time and cut free the rabbit's head. I hold it deep within the palm of my glove and offer it to him. He stops feeding, looks at it, wants what I have, thinks it's a fair swap. He hops off the rabbit and bounces across the floor on to the glove. He continues to tear and pull at the flesh on the skull. It cracks and splits under the strength and power of his beak. Fragments of bone around the eye socket collapse, he pops an eyeball and fluid spurts out on to my shirt. I kneel down and slip what remains of the rabbit into the pocket of my hawking jacket. After twenty minutes of tearing and peeling the skull, CC's crop is fully distended and he stands sated in blood-soaked contentment on the glove. His vocalization is reduced to a low, grinding beep. He leans low over the glove, cleans the dry, flaking meat and blood from the side of his beak along the leather edge. He once again leans over his shoulders wiping fur from his eyes. His grip lessens and we begin the long walk home. The bumping weight of the dead rabbit pulls the straps of my jacket into my shoulders. As always, and with deep-seated seriousness, I am aware of the brutality of my actions and the responsibility I owe to life taken.

I have eaten rabbit hundreds of times before. When I had money to waste I happily paid for the pleasure in restaurants. The rabbit, presumably farmed, had passed through many hands before arriving at my table. The distance between its life and my food absolved all concern about the way it died. I think of the television shows that have reduced the most intimate parts of once magnificent and vivacious animals to mere props. I remember my 5 a.m. winter starts working on the mechanical lines of a slaughterhouse where for six months I participated and witnessed the killing of 250 cattle and many more pigs and chickens in a single day, a level of processing and production so fast and vast that a certain type of complacency kicked in. It was a place where pieces of offal, heart and chunks of lungs were thrown for fun between lines. Where blood was flicked or smeared in faces as a way to cure boredom and elicit humour. A place where wages were low and the context of mass production turned animals into humiliated units sliced and diced without thought or effort, and where the final product was delivered to the five main supermarkets in Britain, the rest frozen in a huge mountain of meat never to be eaten.

I feel the rabbit's blood drying tight on my hands and under my nails. I accept full responsibility for its life and for every other life we intend to take during the rest of the season. I feel comfortable with the way it died. I sense the authenticity of the moment. I understand the hard work we have gone through to attain it and I feel the direct honesty of my actions and those of the hawk. There is no moral ambiguity. The rabbit was hunted and died naturally, was killed fairly and with respect. Better still, many of its brothers and sisters escaped back into the landscape to continue living.

Back at the cottage, I place CC in his weathering and go indoors. I cut the rabbit up, remove the bones, divide the kill, feed some to the dogs, then cook the rest quickly over an open fire. My stomach distends in delight.

The following morning CC still has rabbit left in his crop, so we rest, returning to the field the following dawn.

Trying to get my son to eat is something I find far more difficult than giving him a bath. His taste buds are exceptionally sensitive, in much the same way as my eyes are sensitive to light. I once sneakily fed him bread on the edge of being out of date, and he spotted it instantly. What he likes to eat can therefore be complicated. His mother leaves me instructions, but I forget. He is skinny and grazes, eats like a sparrow: peck, peck, peck. I am worried he will get hungry.

Left to my own devices, I also fall into routines around food, often for months on end. I drink cold tea by the pint mug. I once ate so much pork belly I ballooned up to fourteen stone. Then I switched solely to cucumbers. Whatever I do eat, it is usually only once a day. My feeding habits are perfect for falconry: I can eat any number of pheasant or rabbit without the need for choice or variation. My son is the same. He has a broad range of food that he eats, he is willing to try different things, but he has favourites that he returns to time and again, all prepared to an exacting standard. A meat-and-veg, keep-it-simple type of person.

I go through a list of breakfast-style food. He normally has a jam sandwich but he is not hungry. I open the fridge and freezer and look for ideas. He ends up having a big bowl of vanilla ice cream, followed by another, with golden syrup on it. He thinks this is a marvellous turn of events and begins bouncing off the walls within fifteen minutes.

It is the worst parental decision I have so far made.

I watch CC leave the glove, and the rabbit runs across the grass, twisting and traversing along the base of a fence. CC hits the wire with full force. His neck snaps back, folding like a punched cardboard tube. He falls to the floor, sprawled in leaf litter. He vibrates and begins a weird, floating, rotating fit. His head flicks left then right. He stands, he falls, staggers about, cannot focus, looks drunk.

He is dying.

Rooted to the spot in shock, I watch the last jerking shudders of his life. Searching for safety, for solace, he rows pathetically across the floor back towards me. I experience a heartbeat pause of unshackled pointless drift and detachment. It feels as if someone has stroked an electrode over the soft surface of my brain. I wretch and taste vomit, shout out loud. My senses close down. Standing in silence, I watch him unwind and finally stop moving.

Two or three seconds later he twitches. Then flaps. Then rolls over, stands and screams. I scoop him up and hold him tight. I swear at him: 'You bastard sod!'

As he slowly comes round and regains the glove, I descend into a foul, angry mood. I curse the stupidity of the private worlds and private land of England, where a lot of goshawks die hitting fences. I am not a romantic. There has never been a halcyon day when a goshawk was free to fly any-where in the United Kingdom. *Kingdom*. I curse the word. I wish a pox on all their houses, I spit on the past. I curse those who started carving up the land for their own selfish gain. Enclosure: a process that has nearly killed my hawk. Above all else, I curse myself for letting CC go. I should have held back.

I know this is just the start. A heavy concussion does not always kill a hawk. It can take several days for blood to seep into the brain, causing an embolism, the trauma and shock of which then kills them. Even if this does not happen, if he was to have another knock of this magnitude he will die instantly rather than be knocked unconscious. What is worse, there is also a high probability of a secondary infection, the vilest of which would be aspergillosis.

Aspergillosis is a tiny fungal spore that settles in the internal cavity of a hawk. Birds do not have a diaphragm; their bodies are exposed to infection in any part of their chest, stomach, lungs, trachea and even their wings. It is warm and damp in these little pockets of space and the aspergillosis spores can settle anywhere and multiply. It is a deadly disease and tough. It hangs about, takes its time and is powerful enough to kill a human. Eighty per cent of goshawks suffering from aspergillosis die.

Bill Sykes was a resplendent male in mature plumage and came to me on loan from another falconer. Like Boy, his eyes were the coal-fire orange of maturity. His feathers were a slate grey and blue. Kept locked away in a mews for over four years, he was a difficult hawk to handle or even understand. Bill's silent, locked-down, parent-reared moods were complex. Like a recidivist sociopathic prisoner on day release, he just could not settle.

The first prey he caught was a grey squirrel which bit him hard above his ankle. The hole, positioned badly, was in an impossible place to stitch. Left to heal naturally, the vet pre-scribed a huge dose of antibiotics to counteract potential infection. Giving antibiotics to a bird of prey lowers its immune system. It was Hobson's choice, a balance between treating the immediate problem or opening the door to other infections.

After a few days I detected a slow, low rattling: a thick phlegm-flecked gurgle in Bill's breathing, like the sound of mucus-covered bubbles blowing up through a drain. At a specialist surgery for avian medicine, a circular endoscope was pushed down Bill's throat. The images bouncing up the optic fibre showed an internal lunar landscape of undulating pale organic surfaces, ripped, broken and slit by a dense fungal growth curving and folding like mould back over his trachea. Deeper down, scarred and bruised lung tissue revealed a previously unknown infection. The spores of this recent infection had set deep on the lip of Bill's air sacs and his chances of survival were slim.

Treatment started with the extremely expensive surgical removal of the aspergillosis spores, followed by three types of complicated medication, to be administered over the next two months. In order to speed up his recuperation, I fed Bill up to his highest, healthiest weight. Consequently, he was not hungry and became highly agitated. Two of the three treatments required crushed pills and the secretion of a liquid inside his food twice a day. I would dose him with medication, only to return an hour later and, frustratingly, find a foaming pink pill under the perch or liquid flicked up across the wall.

Taking the medication every day was critical to his health, so I took to physically forcing the medicine into him by hand. He hated me with a vengeance.

A third, concurrent treatment required the nebulization of an avian disinfectant called F10. A nebulizer is a small, noisy pump and tubing system used to treat mouth, lung and throat disease in humans. A mixture of water and F10 flows through the nebulizer and explodes in a thin mist which is breathed in through a face mask and deep into the lungs. For it to work on Bill, he had to be placed into a large box with the door closed and a tube inserted into the side.

The nebulization took half an hour twice a day, once in the morning and then once again in the evening. Bill frantically beat at the door, trying to escape, smashing and breaking his feathers. He bruised the flesh over his beak and cut his face. This treatment went on for more than 200 hours.

The prolonged procedure tested my faith in falconry and my love of birds of prey and nature itself. Ultimately, I saved his life, but Bill would never hunt again. Many, many times, I felt it would have been kinder to let him die.

Waiting for infection to strike in CC, I sink into darker reflections motivated by the fear of his mortality. I start to see the strange contradiction between falconry and our relationship to death. At one end, I actively seek its presence in the quarry, at the other I resist it vociferously, trying to protect the hawk from accidents, illness and injury. The distinction

I make between the two is motivated by a false dichotomy between the hawk's life and that of other animals.

I know from watching CC flailing about on the floor, seeing the aspergillosis spores in Bill, the peel and poke of Girl's frayed talons, Boy's broken feathers, that nature does not share my distinction. It has no such hierarchies. Life throws death about in hard-hitting moments. Nothing escapes. From top to bottom, human to hawk, rabbit to pheasant, flicking insects, soft slugs, nematodes, plants, down deep to microbes, at all levels and at all times we are connected by two simple things: the vitality of our lives and the certitude of our death.

In the short time I have shared watching CC grow, seeing him breathe, feed, fly, feeling him break into my body, the time we have spent being 'out there', out flying, I have become deeply connected to him. If it was a case of a talon or broken feather, I would not hesitate to intervene. But aspergillosis is a wholly different matter. His life, like any other, should be lived as a galloping, free-wheeling experience. It should be lived through the immediate free expression of positive natural behaviours. What is his life worth, if he is broken and left standing on a perch for the next fifteen years?

The pain of this realization becomes too much. I am filled with deep sadness. I am powerless. Instead of fighting it, I draw myself down into it and let it roll over me. I think maybe I should turn my back on falconry and stop. That I should

absolve myself and bow out. Stop participating, and walk back
to the ease, expediency and security of a life without hawks.
But the whole process, the totality of hawks and hunting, are
intrinsic to my identity. Cut me, and I bleed feathers. It is the
closest thing to what I consider sacred. It would be like stop-
ping breathing. It would be like denying nature itself.

As a respite from the scar-tissue toughness of goshawking,
to see if I am all right, Steve and Hollie come to visit with
CC's stepbrother, Sir Percival the merlin. He is the diametric
opposite of CC on many levels and the break is a welcome
distraction.

When he was laid, Sir Percival's egg was small with a light
copper-brown base and splotchy dark chocolate patches rolling around and down over the top. When hatched, he was
the size of a bumblebee, his beak no bigger than the yellow
piece of pie from a Trivial Pursuit boardgame. His legs and
feet were pale pink, a pair of thin, stringy worms that have
toughened and turned yellow over time.

Steve places him on a post perch in the front garden. Sir P
is small and slight, ten inches tall and five ounces in weight.
Picking him up, he scrabbles about on Steve's bare hand then
stands on the table in the cottage. He preens and conducts
himself with a self-possessed bravery. He has an air of arrogant
intensity, is inquisitive and cheeky, a spring-released, fearless,
bouncing mousetrap. He reminds me of a charming child,
a son, scoffing handfuls of Haribo for breakfast. Coated in

a fine, lustrous, mustardy rust brown, unlike CC, Sir P has evolved to fly small birds with persistence and determination.

Distil the purest, finest, most ethereal points of falconry, condense them into a feathered entity, and the result is a Sir Percival. If they were any bigger, merlins would be the greatest bird of prey in the world. Steve tells me that, when Sir P flies, he is as brave and persistent as a gyr falcon. There is only one bird that is an even match for the concentrated supremacy and mighty-mouse power of Sir Percival.

A ringing, singing, ascending skylark rises high over summer crops, drops, dives; a white-edged tail spreads before it mounts once more, hundreds of feet. It is a feat of remarkable endurance and style. Up with the lark and larking about, a merlin will match the flight of this little songbird equally. Flying a merlin is the only branch of falconry not overly concerned with catching. The flight of Sir Percival is one of aesthetics – a work of art, refined in fine lines drawn across a late-summer sky, writ large in duration, context and form and controlled under strict licence for six weeks of the year. A merlin and its prey embody perfect, pure flight without the need for blood. This I what I need. This is a good thing.

We load the car with the dogs and drive from the cottage for several hours to the hawking grounds. Larks nest in little scuffs no deeper than a heel mark in the ground, laying thin-shelled eggs the size of a penny piece. In this area of England during the spring I counted several nesting pairs with three or four eggs to each. The larks had a good breeding season, as they did last year, the year before and the year before that. Their habitat and overwintering grounds were key to their survival rates and numbers, their fecundity an example of careful and thoughtful farming.

In November they grouped together in large flocks, rippling out across the stubble, flickering wings, dancing, drifting up high in the strongest winds, leaving me spellbound. How such a little bird creates so much energy and stamina on a diet of seeds and small insects is nothing short of miraculous.

Sweating in T-shirts under a mid-September sun, Steve and I walk Sir P across the stubble. A light breeze brushes our skin, cools the heat, the world pin-sharp bright, the dry ground shimmering and wobbling in localized mirages. The horizon is a pale cerulean blue that rises overhead into the cleanest ultramarine. Birds are high in the sky and spread across adjacent fields, thirty or more, invisible, save for the most melodic song raining down through the heat in complex golden notes and overlapping sound that spreads for well over a hundred acres.

The first few flights fail. Sir P notes some strength in the larks and pulls off shortly after flying them. Each time, he glides back around and lands on my head, stamps around in frustration, pulling at my hair, his talons like acupuncture on my scalp. Finally, on the crest of a large, rolling field, the right lark fires into the sky. The merlin, on a mercurial, mechanical wingbeat, quickly locks in behind it. Pacing one another, they skirt up in concentric circles, a cone-shaped twirl, and roll around the sky like the pearlescent sweeping spiral of a sea shell. Two delicate tango dancers, they joust, push and parry one another back and forth as they climb. On the unfolding line of flight they continue to go up and up, like blown leaves spinning on thermals. The higher they rise, the louder the lark sounds; it twists in the air, the sun shimmers off her pale feathers, she is confident, cocky and mocking. They curve past the sun. I squint, close my eyes. Their flight line flashes white on the back of my eyelids. I open my eyes and try to refocus. When I see them they have slowed, almost stalled, and are seemingly motionless at a height of 200 feet. On the cusp of contact the lark stops, drops, and Sir P stoops. Twisting as they fall, the lark dodges, skipping away as Sir P tries snatching at speed. The lark slips into the crack of a haystack. Percy pulls up, skims low over dusty stubble and returns to the lure swung by Steve. The flight was poetry in motion and is enough for today. We feed Percy his full reward and head home.

At the cottage Steve and Hollie are less rigid than I am. They have brought a huge feast of five-topping pizzas, pork ribs, chicken, little bottles of French beer, chips and dips. I break easily and tuck in. It would be rude to decline. Before leaving, Steve declares CC to be free of the shadow of secondary infection, free from aspergillosis and simply says:

'Just remember, they are working animals, not pets. This

isn't Kes, and you're not Billy Casper. Accidents happen. If you can't accept it, buy a sheep and hunt grass.'

When the time comes and we step back into the field, CC returns with unstrained vigour, lust, style and surprising intelligence.

My son comes down from sleeping distressed and pale. He has been vomiting. He is wan, yellow and has dark bags under his eyes. I have seen lots of sick children and sent them home from school without a second thought, but the feeling that spirals up through me is very different. It is horrible. The secret spirit that drives my son has been sucked from him. This is a side of him I do not recognize. It is awful. I feel protective but also powerlessly inept. His illness remains and, before he recovers, I have to leave without seeing him back to full health.

He continues to be ill for several days and when I call and speak to his mother she tells me that, although he tried his best, he missed his caveman day. He did not get to wear his costume to school. He was extremely upset not to be a chief catcher. I do not care. I am happy that he is eating and drinking fluids and not throwing up. I tell her that when I next visit I will show him how to make fire using sticks and we will film it so he can take it in and show his teachers.

Having nearly died trying to catch a rabbit, by association, CC begins to hesitate when slipped at further rabbits. For several days I watch him work through the experience of hitting the fence. On the surface, he seems to be flying with increased determination, launching off the glove towards any rabbits at distances of three or four hundred yards. At the moment of contact the rabbits miraculously escape. I cannot figure out what is wrong. Eventually, I get close enough to see what is happening.

As we move round a hedge CC lets out a scream. Hearing his call, a rabbit ducks down in the centre of the field, ears flat, perfectly still. Viewing the situation, waiting to make a move, the rabbit twitches forward, then stops. It is enough. CC moves with speed. The rabbit, overwhelmed, stays still and flattens even further into the grass. At the split second of contact, CC flares out his tail, stalls mid-air, pulls his legs up and bounces back as if on an invisible cord, then peels off and lands in a nearby tree. The rabbit takes full advantage, hippity-hops and ambles across the remainder of the field, through a fence and into the safety of his burrow.

I have my answer.

To continue to hunt would be pointless and would only compound the problem. In CC's mind the association between rabbits and injury has been established. He is pre-empting hitting a fence even when it is not there. As with attaching his backpack and tail mount, he needs time and motivation to forget. For the next three days I take a tip from the Hiebelers' methods. I feed him as much food as he can eat from inside a split-open rabbit carcass. On the fourth day I feed him nothing. On the morning of the fifth day he kills his second rabbit within five yards of leaving the fist. The association is broken.

We slide into another blank patch of over a week, and CC's behaviour switches back to frustration and, in particular, screaming. Again, I try to figure out why. When I leave him on his perch in the garden and take the dogs for a walk, his noise carries clearly across several fields. The rest of the natural world reacts accordingly. The problem of our continued failure is now less about CC's ability and more about the adaptive ability of his quarry. The rabbits have learnt our routine, know our noise and slowly dissolve away from their usual haunts. I make a switch from hawking in the afternoons to the early mornings.

A summer dawn at 4 a.m. is a procession of vast tangerine fire skies and mist rolling over grass, a perfect world, vibrating, full of potential. The unrestrained natural noise released by lack of human activity allows all wildlife to move about at peak freedom and in abundance. Summer dawns feel like a momentary piece of prehistory in a modern world, an echo of what the countryside was before we multiplied. For several weeks we exist alone and the early-morning sessions are beautiful periods. More importantly we begin to kill again.

CC's fitness, determination and success reach peak levels. Wisps of mist swirl around the outer edges of his wings as he slips off the glove, repeatedly hurtling through the dawn. On several occasions he teases me, makes my heart stop dead. I watch him fold his wings, slip through the square metal netting of the sheep fences at over fifty miles an hour. He is so close, and more than once his tail bell pings on galvanized wire. In these moments I fear for him and love him in equal measure. He reminds me of a small boy showing off by pulling wheelies or riding a ramp of his own creation. I swear and shout encouragement. He repays me in kind. He never hits another fence.

It is early morning. I am silently standing outside CC's mews. The silvery darkness of this dawn has the texture of a total eclipse of the sun. My breathing is shallow and excited. In the hoary half-light the silhouette of a wild female goshawk is perched on an electricity pole twenty feet above my head – she is right above CC's aviary. This is the first wild goshawk I have ever seen in England. Presumably attracted by his daylight calling, she is huge, easily three pounds in weight. I open the mews and prepare CC. She is still there when we step into the stubble. CC notices her, tilts his head at an angle and becomes silent. She heaves out of the branches, away to the distant treeline twice as fast as CC.

Hunting out of the dark dawn into the daylight, I walk him from the cottage, exploring different spaces and places to keep the food coming. We arrive at a disused sand quarry. Cresting the ridge and pushing through a row of larch, the ground drops sharply, a treacherous incline down to the base of a bowl a hundred and fifty feet below. The hole in the ground is about four acres wide. Two decades previously, the pit and its fine sand were scooped out with giant mechanical hands. The bounty plundered and sold, it now possesses beautiful neglect, a secret cove known only to a few locals. The untouched, delicate sparseness, bleached dry roots and fences crumple at odd angles; there are unexpected creatures in unexpected places. Moss and small tufts of grass, bramble and lichen grow across the earth. I have seen snakes and slow worms, hundreds of different insects, and like the desert scrub in Texas, this place contains a vast number of rabbits.

As we move about on the lip of the quarry the lunar-like surface is illuminated by the blood red and the pink coral of the morning dawn. Long shadows are cast by mounds and bumps, a surreal Saturn landscape. On a soft sandbank I stop to watch a weasel hopping out from under a thick tumble of dead branches. CC looks at it. He has never seen one before and its strange movement makes him silent. Dizzily focused, the weasel's attention is elsewhere and the wind carrying our scent is in the wrong direction. A twitch of his tail, and he zips down a rabbit-hole ten feet from us. We wait.

The building tension is almost too much. A light drizzle descends and a double rainbow materializes over the crater. I hear a sudden rumble of thunder, a thumping sound from underground. A pile of early-autumn leaves bulges at my feet then explodes as a rabbit bolts free, fleeing from the weasel. CC makes a short, powerful flight and pins the rabbit to the floor. When CC is settled on the glove eating the rabbit's head, I cut the back legs off the carcass and place them in the woodpile. We find a quiet spot and sit on the damp ground waiting for our furry hunting companion to rejoin us.

In the distance, an unfamiliar noise breaks our silence. The morning commute, cars hissing along the road through a warm morning shower, a demarcation and the outer edge of a wholly different world to the one I exist in this morning. I remember it well, and I know which I want to belong to today.

Autumn

On clear, chilly nights there is no light pollution here. The moon and stars are free to appear in glorious multitudes over the cottage. Directly opposite my front door is a huge oak tree. Cloaked in darkness, it takes on the shape and silhouette of W. C. Fields: a round, bulbous head with swelling edges, bumps and a bobbly, nose-like protuberance. Tilted, the constellation of the plough sits suspended close to the top edge of the canopy. When it is in this position I know autumn is approaching. The soft scent of rotting apples and mouldy leaves wafts down the path. The empty frames that held runner beans have been blown by the winds and stagger left and right, empty and broken like old scarecrows. The smoke from the fire is noticeably heavier. As the temperature drops it blows down in thick clouds, over the hearth and into the cottage. The shortening days bring the first tentative blossoms of frost.

A fit goshawk will kill rabbits all day long. Once, when masculine competition and goading surpassed sensible behaviour, Steve caught nine in a short afternoon. If so inclined, he could have caught a lot more, but they proved too heavy to carry from the hill. CC is now so adept, confident and clever that he could catch eight or even ten rabbits in a day if he was allowed. But if this was the only point, I would buy a gun.

Instead, it is time to leave them alone and move with the season, move to what is there to harvest. Feather on feather, flight against flight, feather is the measure – it is time to hunt a different type of animal. England used to be awash with wild indigenous quarry. The famed Elizabethan goshawker Edmund Bert once caught a dozen brace of grey partridge in a day. In the 300 years since Bert walked his goshawk across the landscape the numbers of wild quarry and natural spaces are very much depleted. Much as I would love to replicate Bert's achievements there are not enough grey partridge in my area to sustain CC for a day, let alone a season. So we turn instead to a released bird common to many.

I heard the first pheasant a week ago, a rusted, grating, kek-king call in the early morning. More and more are beginning to appear: I hear them in the distant woods and find them pecking at dawn on the spilled grain in the garden and up the lane. Pheasants, duck and migratory birds are the truest test of any goshawk. More importantly, these birds provide higher nutritional balance and higher quantities of meat than a rabbit: two important natural factors for both of us when the winter arrives.

A heavy thunderstorm the day before has drenched the ground and the dawn is wet and patterned with fallen leaves. Columns and sheets of mist hang heavy, the beams of sunlight catching particles of moisture in the air. The stubble and smashed stalks of the rapeseed are soft, soaked with beads of water.

Two days ago CC killed a pigeon, his first feathered quarry. Just a youngster, it took flight only for a few yards before being caught in the air. It was important to let him eat as much as he could and cement a positive association between food and creatures that fly. Having rested for the day, he is overweight, but his instincts have not been directed or fulfilled. He now stands ready. The feathers on the back of his head are crested.

His focus is intense. His grip is slight and his head bobs forward; his calling drops to two or three screams every ten minutes. He is in full *yarak*. This is the first time I have seen him at this point.

We enter a potato field with long, steep piles of soil under the dying, brown plants. Flash runs ahead thirty or forty feet, leaving Etta to potter about and struggle with the irregular earth. I turn to look at her just as she pulls up, slowly freezing on point.

Instinctively, I tell Flash to 'stay'; he tries his best, but he knows what Etta is doing and keeps bum-shuffling a few inches forward as the seconds tick by. I step towards Etta. Before the command to flush has left my lips, Flash is off into the cover. The first pheasant of the season launches out from under the green. Flash jumps, trying to snap it from the air; CC jinks out of the way then rights his flight line. The pheasant rises higher and higher. CC follows with a quick, clipping wingbeat. Together they mount over the trees and stream and disappear.

It takes time running in the same direction, pushing through heavy hazel and bramble and dropping down into the water to climb up the opposite bank. The dogs begin barking at the fence. I go back and lift Etta and Flash over, retrace my footsteps and run along the top of a far bank. I stop and listen. The new dawn is muted, all sound woollen and muffled by mist. I hear no bell.

I scan about with the telemetry, picking up a signal 200

yards further down the bank, back to the stream, back the way we have just come. As the signal gets stronger, I hear the tin-scraping scream of CC's call somewhere within the trees. The sound is difficult to pinpoint. It seems to drop lower and lower towards the ground. I jump a small gravel bank in the centre of the stream and nearly step on him. His wings are splayed across a branch in the water, the rest of his body submerged up to the chest. The cold current turns and buffets him tight into the curve of the far bank. I walk into the water and feel beneath him, searching for his feet and jesses. My hand brushes wet feathers in scaly feet. The bulge underneath is undeniably his first hen pheasant.

Feeding him on dry land, it takes half an hour of concentrated gorging before he hits the full mark and starts falling forward with the weight of meat in his crop. I do not have to offer him a reward; he steps off the kill, presumably sick of the sight of pheasant. There is not much left. Etta snuffles about and devilishly sneaks off with the head, crunching it gamely just beyond reach. Flash is sitting next to me, shivering in anticipation, so I give him two chicks to eat. When CC is on the glove I trace the feathers back to the point of contact. He caught it on land and clearly dragged it to the water and drowned it. His father did the same thing countless times.

We take a satisfied, slow, circuitous route home, down a lane, threading through a graveyard near the church. In the distance, in the middle of a cricket pitch, a small white sphere breaks up the green space surrounding the roped circle of the crease. I head straight towards it. I have been watching the puffball mushroom grow for three days. I noticed it when it was the size of a tennis ball and it has slowly expanded into a giant, foam-filled bubble reminiscent of a football. It needs to be picked before anyone else finds it. With my free hand I gently roll it off its roots. Flash and Etta sniff suspiciously, bat it with their paws and try to chase it like a ball. CC simply glares at it. I have to cut it in half to get it into my jacket.

When everyone else is fed and settled, using a long-bladed bread knife I cut the mushroom into thick two-inch strips and fry them; they're the size of a T-bone steak Desperate Dan might eat. It has the texture of marshmallow, and browns to a strange, glowing lemon yellow. When ready, it has the smell and taste of warm air blowing through woodland. I pile threads of cooked rabbit and pheasant on top and, like CC, I gorge myself stupid.

As we head towards winter all the lessons and misses, the scratches and bumps and the killing have toughened CC. Psychologically, he knows he can kill anything that moves, and his demeanour and muscles are of a different calibre. His aggression, now focused outwards on to the quarry, has tapered off towards me. I no longer need to use the lure. He returns to the glove from hundreds of yards away and lands without threat. His relentless screaming is much reduced. He instead offers up more intimate noises: occasional chupping, slight hissing beeps and tweets arrive in an assortment of volumes and tones. By turns when out in the field, he changes shape, at times folding his wings tight like a thin spear or puffing up, and rouses relaxed when on my fist. He freely preens and cleans on the glove as we walk the fields. His feet express moods also. No longer showing deep anger on the scales, he can be weighed without any problems. When cropped up and fed, he reverts to his juvenile behaviour, delicately nibbles the tips of my fingers, pulling and preening my hair or the fur of the dogs when they're on the floor.

The dogs have also changed. Etta and Flash rest more. The muscles on their legs break out in angular mounds when they move. Etta, in particular, has veins bulging across her legs and chest. The coats of both have a sheen and lustre that matches CC's feathers. They are so motivated I need only touch the telemetry and they begin to whine and vibrate. They can smell the adrenaline and drive to be out; they have an itchy restlessness in much the same way as CC does when moving into *yarak*.

I too have undergone physical change. I am not a strong person. I am six feet tall but do not exude any trace of masculine presence. I am thin, feline, feminine in comparison to most men. I drop from eleven stone to around nine. I catch sight of myself naked in the bathroom mirror and immediately think of the shape and smoothness of my son when he swims, or when we have our 'jazz band' bath time. I am fascinated by our linked difference. I find it strange to think I was once like him but now have become something else.

I have two cuts on my eyelids and a slash on my nose from blackthorn. I am weatherbeaten, scarred; my eyes are clear but red rimmed and puffy from tiredness. I have the beginnings of my father's nose. My lips are chapped and flaky and I have deep crow's feet on the edges of my eyes. The flesh beneath my collar line is translucent and pale. I see the threading veins beneath the skin, can almost see my heart beating. My chest hair is patchy, has grey hair popping sporadically in colours reminiscent of the mature feathers of the Haiders' goshawk. Excess fat from my sides and midriff has shrunk. My ribs stick out angularly. Parts of me look old and worn. When I bend, the sinew running in a 'V' across my groin is prominent, like thick cable. When I stretch my legs, the muscles are tight and wiry like those of a long-distance runner. I tense my neck and it becomes string-taut but does not snap back. The skin on the right side of my shoulder, where the straps of the telemetry rest has cut and rubbed, is sore and striped in welts. From below my waist and running down around my calves scabs and marks from bramble and blackthorn fight for space with the slashes and dark puncture holes of barbed wire. Bruises, cuts, and scratches on the joints of my hands and feet have split open and have a rough, sore edging. My fingernails have black dirt, black blood and red earth around the cuticles. The heels of my feet are blistered, the base yellow and white with cracked, flaky skin. I see the old nose in wire wool and laugh out loud. I stretch and feel the far edges and deep centrality of my fading body. I see the embodied rough truth of my evolution and life. I know who I am.

'The Ride of the Valkyries' spills out from the radio, which is tuned to Classic FM. My brain snaps back from mapping flesh. In childlike reverie, less Colonel Kurtz, more a dancing amalgam of Charles Hawtrey and Mr Bean, I run through the house, flaunting my nakedness at the dogs.

The horror. The horror.

Winter

Deep winter in the cottage is hard. I feel the extremes of relentless cold. The days are short and there are long, dark hours locked down in the house. I poke my head over the edge of the duvet and look at the window. An old, flaking frame of wood, thin glass. Overnight moisture has condensed to water, from water to ice. The inside pane is frozen, the window a solid sheet of crystals. Flash is at my feet, Etta curled up against my chest and stomach. I am toasty warm and do not want to get out of bed. I watch my breath billow out in front of my face. For a few seconds I try to make steam rings and fail.

I remember I went 'crazy' and bought firelighters, twigs and a bag of coal. If I am quick enough I can get a fire going in less than five minutes. I skip down the stairs, set the wood alight, run back to bed and wait. The dogs, lazy and cunning, have remained under the blankets. I look at the clock: it is 7.38 a.m. The window starts to glow a chalky, opaque orange, like the skin on a segment of tangerine. From the far end of the garden I hear CC calling from his mews. This is unusual. I worry if the cold snap has dropped him down in weight.

We have stepped out of synch with a lot of pheasants recently. These are the ones that have been too strong or fly too high to be shot and have left the sanctuary of feeders and the warmth of the wood to strike out on their own. These pheasants have survived, are smart, fit and take flight easily. They move and act almost as if they are a different species. CC has taken his time to recalibrate; he is learning about them all over again. There has also been a long warm spell when it should be cold. I saw butterflies and bluebottles. The unusual temperature has thrown CC out of condition. He has switched off, so I am forced to be patient. With every missed flight I am a step closer to relying on eggs, fish and vegetables. I make porridge in bulk and consume it with what remains of the apples from the trees in the garden. Kindly neighbours start to supply cake, leftovers and other generous treats from early Christmas parties and soirées.

Some days I lose heart. It begins to feel serious, a bit boring and a grind. As the cold weather returns I count the lumps of coal, ration the small scoops I put on the fire and wear a coat indoors. When we return from the field and have failed, I swing between frustration and then, re-doubling my efforts, force myself to remain focused. The feeling of freedom ebbs and flows and sometimes I suffer psychological exhaustion. I question the validity of my actions, scorn myself for even trying.

I begin to stop and pick up the carcasses of dead pheasants that litter the small roads and lanes. This does not bother me; I have done it intermittently for most of my life. The kills by cars surpass CC's by a ratio of roughly two to one. To ignore the free food would be a waste. It is impossible to keep up and consume them fresh, so I butcher the breast and legs, make stews, curries or pies and freeze them. I am grateful for people's inability to swerve their cars.

My patience is finally rewarded. CC takes a fluke pheasant against a fence. His success builds incrementally. He takes another near the sandpit, then one near a nature reserve. On each kill I gorge him then do not feed him the following day. He takes another, then another and then a rabbit. A switch flicks as the frost and snow return. The cold triggers a change and we move up to the final level: our hunting bubble truly returns. I increase his weight, a quarter-ounce, half an ounce, one, two, three, four ounces, in a short space of time. I no longer have to weigh him. We are flying on routine, appetite and not hunger. If I did not feed him for a week, he would still be alive. He is as close to his natural counterpart as it is possible to achieve, a fully realized goshawk, what falconers call a 'made hawk'. Everything outside of CC and our hunting ceases to be of concern. Our success removes the exhaustion; when he flies he moves and behaves with such beauty, form and grace that all else fades and seems trivial by comparison. The normal dimensions and framework of my reality shifts. We step out of time, the names of days disappear, are once again forgotten. All else is defined by his urge to hunt. I feel half formed and fallow until he is ready. In the field we are locked in deep symbiosis and I feel and see myself moving as he flies. The connection is powerful, so much so that, skimming on mid-sleep in the early dawns, I begin to dream in grey, in browns and the caramel colours of a hawk. When I replay the flights of yesterday, they morph into the imagined ones ahead. Abstract feather forms move through the REM of light sleep and the non-linear moments of my unconscious-ness conflate with lived reality. For a brief period these two states are only slightly out of synch, the slippage minimal. My dreams have never taken this form before or since. The feel-ing is formidable.

To the general population the flight of a hawk chasing quarry looks the same time after time. But every slip, no matter how short, or even if they fail, has its own sense of internal poetry. The canvas and context of sky is different. The lines and movement of predator and prey draw differing shapes and forms, creating an improvised moment that can never be repeated. Against the background of the typical days, the really good flights, those measured in distance and duration, happen unexpectedly. When they arrive, the moment the hawk leaves the fist and the wingbeat is seen, is when the art begins.

Some paintings are better than others.

I begin climbing a steep hill near the sandpit. The morning is cold, the ground solid frost. The sun is high, the air thin, the sky brittle and clear. The horse in the field seems to be smoking: steam is rising fast from her flanks and blanket. I leave weaving footprints of green surrounded by a white coating of frost. Next to my boot, the trident-shaped markings of a group of pheasants appear halfway up the steep incline. Flash catches the scent and runs wildly up and down the far fence. It takes another minute of hard walking before I reach the top of the hill. Flash is into the trees and barking. The wings of several pheasants clatter through the pine and larch to my left. CC hears them and bates to be let go then swings back around and regains the fist. I hold him back and wait. I remove a small camera from my inside pocket and begin filming.

A hundred yards away a hen pheasant breaks out over the field to our right at full speed. Flying at forty or fifty miles an hour, she curves round, mounting fast, climbing high up and out towards us. Even at full tilt it takes her three or four seconds to reach our position, and she passes like a comet overhead and slightly to one side. It is an impossible pheasant. CC leaves the glove, turns right and strides into the sky. The hawk and pheasant melt across a huge silver sun and the ground drops away to the valley below. They are at a height of 150 feet or more. CC closes the gap and tail-chases the pheasant in a straight line across the distance of a football pitch in a matter of seconds. They keep going, streaking high over the firmament. With no containment or control, no fences, no boundaries, the only expression is the expansive, colossal movement of pure, free flight. It is stag-geringly beautiful. As they keep moving it develops into easily the biggest, most formidable flight of a goshawk I have seen anywhere in the world. The sensation is like holding light.

In the very far distance, two small dots slowly descend, curving down to the ground and drop out of sight. I think of Haider, Punhal and Ghulam. I think of Salman. I think of Viktor, Craig and the Hiebelers. They would approve. More than anyone, I wish Steve could have seen what the hawk he bred has just done.

I know CC has caught the pheasant. I don't even bother to run. I trot up the crest of the hill, move through a nature reserve, walk down a steep bank, through two gates, across a road and over a bridge. I walk along the hedge of a maize field, tracking the signal on the telemetry. Invisible behind the died-down stubble, I startle a flock of lapwings, well over a hundred birds strong – roughly the same number of kills CC has made. They rise off the land, move from white dots to black bats, ghosting each other in a curving, whip-like motion. The flock bulges and thins, sweeps and dives in silence. Focusing on the head bird, I watch as she moves several inches to the left, the ripple, a knock-on effect like falling dominoes, shift the flock at speed like a shoal of mackerel or a murmuration of starlings. Beyond them, a noisy murder of crows spirals over a ditch next to the road. I head towards them.

I find CC over a quarter of a mile away from the place where he left my glove. He has eaten nearly all the pheasant. I thank him profusely: 'Well done, mate.' There is nothing more I can do. There is nothing more CC can give. I have nothing left to prove. We have made it.

The next day CC weighs well over two pounds. When I pick him up and move him to his perch he is emitting a strange noise, one that I have not heard before. His head bobs upwards and he speaks with low, soft *tuk tuk tuk* chupping. I text Steve and ask him what it is.

'He is telling you he is glad to see you. He LOVES you, mate.'

I love Christmas. I love the time of year. The temperature, the smells, the fact that it is the height of the hawking season. I like the making and swapping of cards and gifts. I like the friendly bon-homie when I pass strangers walking in the countryside. I like the length of the holiday. I love the silly jumpers and log fires. Despite the fact that I am a hardened atheist, 'Silent Night' is one of the most captivating religious songs ever written. It moves me deeply. Unfortunately, I do not like shopping.

I am trying to buy Christmas presents for my son. I have not been into a town or a shopping centre for a long time. Annoyed at bumping into people, I pull my shoulders in and look up. The experience is like being squashed into a cheap striplight designed by Jeff Koons then turned to full power by Philip K. Dick. I am in an alien world, a hyperreal dream, surrounded by the tacky, screaming lights and sounds of a travelling funfair. The Christmas rush around me is immense. I can feel the buzzing tension but not the tension of joyful expectation. There is an underlying aggres-sion, a kind of frustration mingled with the smell of fear. It reminds me of the build-up to a fight in a school playground.

Overwhelmed, my anxiety symmetrically escalates to the time I spend walking the aisles. I have no plan other than to drift aim-lessly and browse. This proves to be a mistake. I become distracted by a whole section dedicated solely to shampoo. Rows and rows, columns in all sorts of flavours, colours and designs. It looks like a clever situationist sculpture designed by Andy Warhol. Unable to cope, I walk back through the shop, placing each item from my basket in the wrong place, and go home.

The other main issue with shopping for presents is that my son has a huge array of toys and games. He has books, technology and clothes. He is not spoilt, far from it, he appreciates all the things that he is given and is fun to take to a toyshop. We usually collect his ideas together in a basket then decide which are his favourites and which can be returned to another time. He is patient, considerate and is able to resist instant gratification. He has all he needs. What he wants is another matter. I decide to attack the shopping with logic and quick movements. I write a list, properly plan my route and get on with it. I win. I find most of the things he wants and add a few more that I think he will like. I know he will love all of them the moment he tears off the wrapping paper then, in a few months, they will be half forgotten. I was exactly the same. I can list on one hand the toys and games I remember from my own childhood. But, like all parents, I would like to give him something substantial, something that will last.

Spring

There is a warmth in the air that I have not felt for several months, and the hawking season will be over soon. Falconry is inherently tied to natural cycles. After the feast comes rest and regeneration. Before spring truly starts, we stop. The animals we seek need to reproduce, grow strong and repopulate the land with next season's food and quarry. This time is also a period of regrowth for the hawks. Spring and summer are for the moult and mating. CC will be released into his mews, fed as much food as he can eat, and his damaged feathers will drop out, be replaced by new, adult ones. I am happy that the end is near. I have enough meat in the freezer and I can soon start fishing again.

My son is about to visit the cottage for the first time. His mother has brought him to see the place where he was born. He is interested in his history. When he arrives I go into overdrive and offer him an assortment of food and drink. He refuses all but the Ferrero Rocher chocolates his mother has bought.

As he mooches through the front door I am a little bit embarrassed at the state of the cottage. He doesn't know the word 'rustic', and the life of a single man is often less than spotless. Maybe he doesn't notice, or even care, but his mother's eyes flick about and scan the small warm room with its low beams and cobwebs. I think she is more concerned about the open fire than anything else.

We all walk into the garden. I untie the leash and remove CC from his perch. My son's mother shares her knowledge of falconry. My son listens intently, discovers a side of her he never knew existed. She has hand-raised and held sparrowhawk chicks, hunted with golden eagles, seen them chasing hares over the dark, fertile soil of Lincolnshire before he was born. It's well out of his frame of reference: he does not associate his mother with falconry and I watch him trying to work out the timeline, trying to work out how this tiny dot of a woman could possibly hold on to a fourteen-pound golden eagle.

I ask him: 'Ready?'

'Yep.'

I slip the dogs from their leads. They skirt wide out into the field adjoining the garden. My son skips and walks ahead at ease. The usual tense focus I feel, the seriousness of falconry, evaporates. He begins chattering away about all sorts of subjects, a free-rolling stream of consciousness. His use of complex words and language always startles me. I ask him if he knows what the word 'mortified' means and what certain other words he uses mean. He gives a clear explanation of each then carries on nattering. I am pleasantly distracted, looking at the hair on the top of his head.

We walk around the edge of a small lake, up a slight incline and through a gap in a hedge. A long way ahead a cock pheasant walks nonchalantly out of the brambles. CC bates. I look up and stop talking. The pheasant stops and almost does a double-take at our weird, marching menagerie. It begins running along the edge of the field. I let CC go. He loops around the side of my son and, for the first time in my life, I am not looking at the hawk. I am watching my son watching the hawk. The sight makes me extraordinarily happy. I look up, the pheasant tries to rise and fly. CC smashes it into the hedge. My son takes to his heels and mud sprays up behind his boots. He runs like me, runs like a falconer.

I overtake him, shouting encouragement, shouting for him to catch up.

I find CC ripping into the cock bird. When my son finally finds me he scrambles between the broken wooden fence and crouches next to me in the hedge. I can hear him breathing heavily. He is so close I smell a waft of chocolate on his breath. I kill the pheasant. Its wings clatter and sweep dead leaves and twigs into the air. CC grips tighter and it stops moving. Peeling the skin and feathers from its chest, I cut it open with my knife. The blood flows freely over the sharp metal and on to my hand. I pass the knife to my son. He is fascinated by the blade. Eyes wide and excited, he watches CC tear into the pheasant, drawing chunks of liver and blood up through a hole in the rib cage.

The head of the pheasant lies discarded in the leaf litter, red, fleshy cheeks against green, shimmering silvers and mottled black. My son picks up a stick and without mercy beats the pheasant's head into the ground. I laugh. His mother chides him: 'Have some respect.'

I stop laughing. She is right. But that only comes with time. I laugh because he reminds me of myself at his age. He reminds me of the tribal children taunting one another with the egret. My son is acting in exactly the same way as them. This is the first kill he has seen and his reaction is not something I have taught him, and it is wholly unexpected. With moral certitude, instinctively he displays an unfettered tactile exploration of the natural world. Without prompting, my son, like the children of the tribe, like children the world over, is simply showing an undiluted interest in food all hunter-gatherers have shown since time immemorial.

At his mother's behest he stops hitting the pheasant's head and inquisitively picks up the long copper-and-gold pulled and plucked tail feathers. Holding them in his hand, he sits back on his haunches, silently sucking up the information spread out before him. Once again, I watch my son watching CC. He is fascinating. His sense of enquiry and ease around the hawk is a revelation. All the travels and torments, all the lost hawks, the learning, the frustrations and failures, of Boy and Girl and CC, flow through me, encapsulated in this moment. It is the most intimate, private part of who I am and, as I cut meat from the carcass and feed CC, I pass the experience forward to my son as an act of love and education. He relishes it more than I could have hoped. He seems to innately understand the vitality of the visceral and, as the life of the pheasant descends, new knowledge in my son slowly rises. He is one of only a small number of children in the world to have successfully hunted with a fully made imprint goshawk. Without knowing it, he has been entered as the most recent participant in a travelling song line stretching back to the ridges of Kazakhstan, to the tribal falconers, and to every falconer from every continent over the last 5,000 years. I am truly proud of him.

Whether he becomes a falconer or not does not matter. What matters is that I have attempted to sow the seed and will continue to sow more. In doing so I have hopefully introduced him to a unique way of understanding the natural world. Whether he allows this seed to grow is up to him. He is smart enough and clever enough to work it out for himself. I trust him to decide in his own time and come to it under his own volition. What is certain is that I want to be there and watch him make this choice. I know I will make many more mistakes, that I will fail and no doubt frustrate him as he transforms into a man. No matter what happens, or how he reacts, I know I will not leave him again. I will always be here in my own peculiar way.

I swap CC off the kill and on to the glove. I ask my son to push the carcass into the pocket of my hawking jacket. We climb over the fence and rejoin his mother and all of us walk slowly back to the cottage. When we get there I notice he has something in his hand.

I ask him what it is.

'This is my blood stick.'

I look closer. True to his word, it is a small stick covered in dark red, congealed pheasant blood. He holds it up to the light. A blood stick as memento and memory. A blood stick as bond.

Blood sticks to the hawk and the human.

Blood ties us both.

EPILOGUE

I am sitting at the back of a room, tucked away in the far-right corner, my back is against the wall and I feel safe. There are about twenty other adults, all of us sitting on slightly uncomfortable benches. The group is made up predominantly of women but there are also one or two men scattered among the crowd.

The children file in through a double door and sit cross-legged on the flaky shellacked parquet flooring. The head teacher asks for quiet, goes through several announcements, they sing some songs and various awards are handed out.

My son is invited out to the front. He stares straight at me with his wide brown eyes and I raise both thumbs and smile. This is my first school sharing assembly. He begins talking and explaining his chosen topic in a clear voice and without nerves. He points to a huge screen and a frozen image of a short video appears. He nods to his friend, who presses play. In the video, the merlin I am holding launches off my glove and flies towards my son, who is swinging a lure halfway across the field. The little falcon zooms *past him, curves up into the sky, holds its position then makes a quick stoop back across the lure. It is a beautiful flight, my son's technique perfect. The collected group of children and adults let out a slight breath of surprise. Appreciation and quiet murmurs begin to flow around the hall. The short film stops and my son explains what they have just seen. He nods again to his friend and a second video begins. This one shows him sitting next to the merlin as it feeds on the lure he has just swung. In the video he begins touching and stroking the falcon, then reaches in and starts pulling off small chunks of meat and hand-feeding them to 'The Phantom'. When he finishes explaining how he has trained his first bird of prey, the head teacher begins clapping.*

As do I.